31973

This book is due for return on or before the last date shown below.

The ~~uem~~

Series Editor

Craig Donnellan

Assistant Editor

Lisa Firth

Independence

Educational Publishers
Cambridge

First published by Independence
PO Box 295
Cambridge CB1 3XP
England

British Library Cataloguing in Publication Data
The Waste Problem – (Issues Series)
I. Donnellan, Craig II. Series
363.7'28

ISBN 1 86168 344 8

Printed in Great Britain
MWL Print Group Ltd

Layout by
Lisa Firth

Cover
The illustration on the front cover is by
Simon Kneebone.

CONTENTS

Introduction

The Waste Problem is the one hundred and eleventh volume in the **Issues** series. The aim of this series is to offer up-to-date information about important issues in our world.

The Waste Problem looks at the issues involved in looking at waste, and how to tackle our waste problem.

The information comes from a wide variety of sources and includes:
Government reports and statistics
Newspaper reports and features
Magazine articles and surveys
Website material
Literature from lobby groups
and charitable organisations.

It is hoped that, as you read about the many aspects of the issues explored in this book, you will critically evaluate the information presented. It is important that you decide whether you are being presented with facts or opinions. Does the writer give a biased or an unbiased report? If an opinion is being expressed, do you agree with the writer?

The Waste Problem offers a useful starting-point for those who need convenient access to information about the many issues involved. However, it is only a starting-point. At the back of the book is a list of organisations which you may want to contact for further information.

*****

The problem with waste

What is waste and why does it matter?

Waste or rubbish is what people throw away because they no longer need it or want it. Almost everything we do creates waste and as a society we are currently producing more waste than ever before. We do this at home and at work. The fact that we produce waste, and get rid of it, matters for the following reasons:

- when something is thrown away we lose the natural resources, the energy and the time which have been used to make the product. The vast majority of resources that we use in manufacturing products and providing services cannot be replaced. The use of these resources cannot go on indefinitely – we would run out.
- when something is thrown away we are putting pressure on the environment's ability to cope – in terms of the additional environmental impacts associated with extracting the new resources, manufacturing and distributing the goods, and in terms of the environmental impacts associated with getting rid of our rubbish.

when something is thrown away we are failing to see it as a resource. It is well understood that what is waste to one person may not be viewed as waste by another. A good example of this is scrap metal which has been recycled for many years. Increasingly people are realising that it makes economic sense as well as environmental sense to use 'waste' rather than just throw it away.

The process of using up the earth's natural resources to make products which we then throw away, sometimes a very short time later, is not 'sustainable' – in other words, it cannot continue indefinitely.

The way in which we consume materials will affect whether we have a sustainable society that leaves resources available for future generations to use. As consumers and producers, we are central to the concept of sustainability. We need to think about how we can use fewer resources ('get more out of less'), how we can make products last for longer (which means we use less and we throw away less) and how we can do

better things with our so-called 'waste' than throw it away. We need to see 'waste' as a 'resource'.

The 'waste hierarchy'

The best way of managing our waste is not to produce it in the first place – waste prevention. After that we can think about reducing the amount of waste we do produce. Then there may be an option to reuse the material. The UK Government has developed this approach to derive a hierarchy of options for managing waste – known as 'the waste hierarchy'.

The waste hierarchy specifies the following order of preference for dealing with our wastes – with those towards the top of the list more desirable than those towards the bottom:

- reduce
- reuse
- recover (recycle, compost, recover energy)
- disposal.

The hierarchy is a guide. It does not mean that in all circumstances, at all times, a higher option will be better than a lower option. In most cases a combination of options for managing the different wastes produced at home and at work will be needed. But the hierarchy provides a simple rule-of-thumb guide to the relative environmental benefits of different options.

The problem we have today is that more of our rubbish is dealt with towards the bottom end of the hierarchy than the top. The challenge is to change our attitudes and our practices so that much more of our waste is dealt with by options towards the top of the hierarchy.

- Information from Waste Watch. See www.wastewatch.org.uk or see page 41 for their address details.

© Waste Watch

What a waste

We live in an increasingly 'throw-away' society

Many things that used to be repaired or reused now become waste. This is particularly true of plastic goods, and it has increased our use of materials and energy as well as creating more waste.

About three million tonnes of plastic waste are produced in the UK each year, much of which is packaging (60 per cent).

> ## Many things that used to be repaired or reused now become waste

Once we have finished with plastic products, most of them are buried in the ground at landfill sites. As plastic is very durable and does not break down, this is where they will remain. So the more plastics we use and throw away, the more ends up in the ground. This is not sustainable, and there are better ways to deal with waste.

What you can do...
- Choose goods without excessive packaging.
- Give away unwanted goods so they can be reused.
- Reuse items such as carrier bags.
- Recycle plastics where possible

Less waste. Resources can be used more efficiently through reducing waste.

More recycling. About seven per cent of plastics are recycled in the UK, but only certain types, and this does not happen everywhere. The technology exists to recycle plastics, but the high cost of collecting and sorting compared with buying new plastic is a limiting factor.

Energy from waste. Plastics have a similar energy content to coal and oil. Energy is recovered from some plastics in household waste at energy-from-waste incinerators and more is likely in future. Emissions from incinerators are tightly controlled by the Agency.

Stop litter and fly-tipping of waste. Most of the litter found on beaches is plastic. This is an eyesore, and it stays around for a long time.

What needs to be done?
Existing and forthcoming European and national legislation, including the Government's Waste Strategy, will help to increase recycling and recovery of all waste (of which plastic makes up less than one per cent by weight, but more by volume). It remains to be seen if these measures are sufficient or if further incentives are required.

The EC Directive on Packaging and Packaging Waste set targets for retailers and other businesses. These were that by June 2001, 50 to 65 per cent of packaging waste by weight is re-covered, 25 to 45 per cent of which must be recycled. And that at least 15 per cent of specific materials, including plastic, is recycled. The UK met these targets by 2002 and the amount of packaging we recycle continues to increase.

> ## About three million tonnes of plastic waste are produced in the UK each year, much of which is packaging (60%)

- The above information is reprinted with kind permission from the Environment Agency. Visit www.environment-agency.gov.uk for more information or see page 41 for address details.

  © Environment Agency

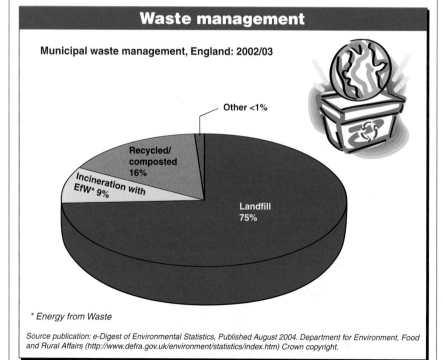

Waste management

Municipal waste management, England: 2002/03

- Other <1%
- Recycled/composted 16%
- Incineration with EfW* 9%
- Landfill 75%

* Energy from Waste

Source publication: e-Digest of Environmental Statistics, Published August 2004. Department for Environment, Food and Rural Affairs (http://www.defra.gov.uk/environment/statistics/index.htm) Crown copyright.

Waste words

A glossary of words connected to waste and the 3Rs

3Rs

The short way of saying 'reduce, reuse and recycle', perhaps the three most important words in the world of waste.

Agenda 21

Agenda 21 is an international agreement to reduce the impact of the human race on the world's environment.

Biodegradable waste

Waste that can break down or rot naturally when attacked by bacteria. Examples include food and garden waste. Other kinds of waste are said to be non-biodegradable.

Bring site

A central collection point where people bring recyclable materials and place them in special containers, such as bottle banks, for recycling. Often found in supermarket car parks.

Compost

Compost is created by the controlled breakdown of biodegradable material such as garden and kitchen waste. It can be used to improve soil structure and nutrient levels without the need for artificial fertilisers and peat-based composts.

Domestic waste

Waste which comes from homes. Also known as household waste.

Energy from waste

This uses the energy contained in waste to generate power and heat while reducing the amount of waste. Examples are incineration used to provide heat to nearby buildings, and methane gas from landfill sites being used to generate electricity. Also known as energy recovery.

Fly-tipping

The illegal dumping of rubbish in unauthorised places such as road-sides.

Greenhouse gas

A gas that absorbs heat and therefore contributes to the warming of the earth's atmosphere (the 'greenhouse effect'). Examples of greenhouse gases include water vapour, carbon dioxide and methane.

Hazardous waste

Waste that is potentially harmful to humans, other living things and the environment and which needs to be carefully disposed of. Examples of hazardous waste include asbestos and poisons. Also called Special Waste.

Incineration

Getting rid of waste by burning it at high temperatures. Around 9% of the UK's household waste is incinerated.

Kerbside recycling

Also known as collect schemes, these are schemes where households put recyclable materials on the roadsides outside their homes, for collection by the local authority or a waste contractor.

Landfill

Most rubbish collected from homes in the UK is buried in large holes in the ground (often old quarries) called landfill sites. Many of our current landfill sites are nearly full and we are rapidly running out of suitable land to create more.

Landfill tax

A tax on every tonne of rubbish sent to landfill sites. The tax is designed to reduce the amount of rubbish sent to landfill sites by increasing the reduction, reuse and recycling of waste.

Leachate

Liquid consisting of a mixture of rainwater and rotten organic materials which drains from a landfill site.

Litter

Waste (usually paper, plastics and glass) thrown around in the environment, rather than being placed in a proper bin or other waste facility. Not all waste is litter but all litter is waste.

Materials reclamation facility

Often known as an MRF (pronounced 'murf'), this is a place where materials are taken to be sorted and stored before they are sent off to be recycled.

Methane

A gas given off by landfill sites and which is both highly inflammable and a major contributor to global warming. As well as being produced by landfill sites, methane is also the main ingredient of the gas which we use to cook and heat our homes.

Mobius loop
A logo meaning that something can either be recycled or that it is made from recycled material.

Natural resources
Substances of use to humans that are derived either from the earth (e.g. coal, oil and metal ores) or from living things.

Litter is waste (usually paper, plastics and glass) thrown around in the environment, rather than being placed in a proper bin or other waste facility

---

Organic waste
Waste derived from plants and animals makes up about 20% of the weight of an average dustbin. A lot of the organic waste created by households consists of food but other sources are garden waste and the contents of babies' nappies. Yuck!

Pollution
Putting poisonous or other harmful substances into the environment.

Raw materials
The basic resources used to make materials and products. For example, raw materials used in the manufacture of steel include iron ore, coal and limestone.

Recycling
Recycling means using things that have already been used, to make new things. This can involve turning the old material into a new version of the same thing. Alternatively, materials can be recycled into something completely different.

Reduction (Reduce)
Reduction means avoiding creating waste in the first place and is an even better thing to do than reusing or recycling. Examples of waste reduction include buying items with less packaging and not replacing items until really necessary.

Refillable
Means that packaging (for example a bottle) can be refilled rather than having to be thrown away when it is empty.

Resources
A general word for the things and materials that we obtain from the earth. Resources can be classified in two ways.
- **Renewable resources** are those that can replace themselves over a fairly short time scale. Examples include the water in a reservoir or crops which grow from year to year.
- **Non-renewable resources** can either never be replaced or take a very long time to replace. Examples include coal and oil.

Reuse
Reusing means using something again, either for the same purpose or for something completely different. Examples include returning milk bottles for refilling and repairing electrical goods when they go wrong instead of throwing them away.

Rubbish
Anything that we think we no longer have a use for and so throw away. Means much the same as waste but not the same as litter.

Sustainable development
This means finding ways to meet the needs of the present generation without damaging the environment or preventing future generations from being able to meet their own needs.

Toxic waste
Waste that is poisonous to humans or other living things.

Waste
Anything that we think we no longer have a use for and so throw away.

Waste hierarchy
This describes the way in which some ways of dealing with waste are better for the environment than others. Reduction of waste is the best option followed by reuse. Only when neither of these is possible, should waste be recycled. Disposal through landfill or incineration should only be the last resort.

Waste transfer station
A place (often a large warehouse) where waste is separated or 'bulked up' before being taken elsewhere for recycling or disposal.

Wormery
A container specially designed to allow worms to break down discarded food and other organic waste and convert it into compost and liquid fertiliser.

Waste minimisation
This refers to the whole process of sending less waste to landfill and incineration by finding ways to reduce, reuse or recycle it.

■ The above information is reprinted with kind permission from Waste Watch's Recycle Zone. For more information please visit www.recyclezone.org.uk or see page 41 for their address details.

© Waste Watch

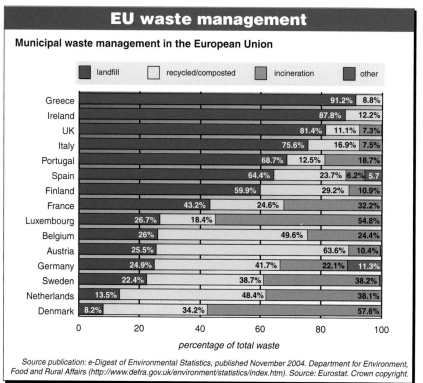

EU waste management

Municipal waste management in the European Union

Legend: landfill | recycled/composted | incineration | other

Country	landfill	recycled/composted	incineration	other
Greece	91.2%			8.8%
Ireland	87.8%			12.2%
UK	81.4%	11.1%		7.3%
Italy	75.6%	16.9%		7.5%
Portugal	68.7%	12.5%		18.7%
Spain	64.4%	23.7%	6.2%	5.7
Finland	59.9%	29.2%		10.9%
France	43.2%	24.6%		32.2%
Luxembourg	26.7%	18.4%		54.8%
Belgium	26%		49.6%	24.4%
Austria	25.5%		63.6%	10.4%
Germany	24.9%		41.7%	22.1%	11.3%
Sweden	22.4%	38.7%		38.2%
Netherlands	13.5%	48.4%		38.1%
Denmark	8.2%	34.2%		57.6%

percentage of total waste (0 to 100)

Source publication: e-Digest of Environmental Statistics, published November 2004. Department for Environment, Food and Rural Affairs (http://www.defra.gov.uk/environment/statistics/index.htm). Source: Eurostat. Crown copyright.

A brief history of rubbish

Information from Recycle More

Millions of years ago before there were people on earth, there was no rubbish. Nature wastes nothing and recycles everything. When plants and animals die, they decompose into the soil, providing nutrients for other plants and animals to live.

When people came along they began to create rubbish. At first, this rubbish was made up of tools, bones and ashes. As the centuries passed, the number of people, technologies and industries increased. Eventually, this led to a large increase in industrial activity called the 'industrial revolution'. By the eighteenth century, factories were making more goods for more people, creating large amounts of rubbish.

Today, there are twice as many people in the world as 100 years ago. New technologies have made life easier and most of us want the latest gadgets, fashions and work-saving appliances. All these things create more and more rubbish.

Most of the rubbish is produced by the countries who are part of the developed world, although they only make up about 5% of the world's population

---

Most of the rubbish is produced by the countries who are part of the developed world, although they only make up about 5% of the world's population. The 'throwaway lifestyle' is not happening in every country – most of the people in the world do not have access to luxury goods such as TVs, CD players, washing machines or cars. Just imagine how much more rubbish would be produced if everyone in the world had the same standard of living.

It is impossible for people in rich, developed countries to carry on enjoying this lifestyle, and to go on buying new products and throwing away old ones. It is important that we increase the amount of rubbish we reuse and recycle, and think about how we can reduce the amount of rubbish produced in the first place.

■ Information from Recycle More. Visit www.recycle-more.co.uk for more information or see page 41 for their address details.

© Recycle More

Waste facts

Information from the Women's Environmental Network

Landfill

■ The volume of waste produced in the UK in one day is enough to fill Trafalgar Square to the top of Nelson's Column.

■ 80% of Britain's populations live within 2km of a landfill site.

■ Leachate is a liquid that seeps from landfill sites. It contains pesticides, solvents and heavy metals and is harmful to health and the environment.

Junk mail

■ The average British household receives 13 items of Direct Mail every four weeks.

■ 32% of it is never even opened.

■ 1 million tonnes of junk mail and magazines are binned every year.

Packaging

■ Every year in the UK, we generate 2.5 million fully grown African elephants' worth of packaging!

■ 150 million plastic carrier bags are used in Britain every week. If you laid them together, they would almost reach twice around the equator!

■ Plastic bags can take up to 500 years to decay in landfill.

Christmas waste

In the UK's post-Christmas rubbish bin, you'll find:

■ 80,000 tonnes of clothes and textiles

■ 125,000 tonnes of plastic packaging

■ 4,200 tonnes of aluminium foil

■ 83 sq km of wrapping paper

■ 1 billion Christmas cards

■ 6 million Christmas trees.

■ The above information is reprinted with kind permission from the Women's Environmental Network. Visit www.wen.org.uk for more information or see page 41 for address details.

© Women's Environmental Network

What happens to the rubbish we produce?

Information from Waste Watch

How much waste we produce and what we do with it are vital if we wish to live in a sustainable society. The reasons behind the processes we employ are discussed in our topics sheet – *The Problem With Waste* [see page 1].

The energy released from burning the rubbish is often used to generate electricity

---

Waste from our homes is generally collected by our local authorities through regular waste collection, or by special collections for recycling. In addition, householders may make special trips to their civic amenity (CA) site, or organise a bulky waste collection in order to dispose of particular items.

Whilst it is difficult to monitor reduction and reuse schemes, councils and waste management companies do collect figures allowing us to note how much of collected waste is intended for recycling (or recovery) and how much for final disposal through landfill. The main methods currently employed are landfilling, recycling, composting and energy from waste plants.

Landfill

At the most basic level landfilling involved placing waste in a hole in the ground and covering it with soil. Today, the engineering of a modern landfill is a complex process, typically involving lining and capping individual 'cells' into which waste is compacted and covered to prevent the escape of polluting liquid or gases. Systems are installed to capture and remove the gases and liquids produced by the rotting rubbish.

Household waste recycling

Recyclate from recycling collections frequently sent to a materials recycling facility (MRF).

At the MRF the materials typically travel along a conveyor belt and the specific fractions are gradually removed. Metals may be extracted using magnets, paper taken off by weight and other screening devises used. Following separation the constituent materials are baled prior to sending to reprocessors.

The activities at these plants are specific to the material being processed – pulping and shredding of paper, granulation of plastics, melting of metals and glass to name but a few.

Many goods produced with recycled content will end up in the shops as ordinary household products, such as bin bags, stationery, furniture, or even filling for duvets and pillows.

Composting

The biodegradable component of municipal waste that will break down is know as BMW and includes kitchen and garden waste, paper, card and more. Composting allows this material to break down and results in the formation of compost that can be used as fertiliser.

Although compost can be made at home, councils are increasingly developing centralised composting schemes for residents' garden waste to tackle this large and problematic part of the waste stream.

Incineration

Incineration is the burning of waste. Incineration may be carried out with or without energy recovery. The energy released from burning the rubbish is often used to generate electricity.

Additional technologies

Research and development of new technologies to deal with the waste we produce are constantly developing. Among the techniques are alternatives to incineration such as pyrolysis and gasification and also anaerobic digestion, mechanical biological treatment and more.

■ Information from Waste Watch. Visit www.wastewatch.org.uk for more or see page 41 for address details.
© *Waste Watch*

Incineration and landfill

Most of our waste is burned or buried which is bad for the environment and our health

Incineration

The combustion of waste at high temperatures:

■ **Encourages more waste**
Incinerators need a minimum of rubbish to operate. To meet demand, local authorities are abandoning recycling and waste reduction plans.

■ **Uses up energy**
Even incinerators that generate electricity aren't an energy-saving option. Recycling saves far more energy because it means making fewer new things from raw materials.

■ **Causes pollution**
Smoke, gases and ash from incinerators can contain harmful dioxins which are a cause of cancer.

Landfill

Dumping rubbish in the ground or in waste mountains:

■ **Releases toxins**
Rotting rubbish emits explosive gases and polluting liquids. Methane emissions contribute to climate change.

■ **Threatens our quality of life**
Landfill creates problems for local communities. Nuisances include more traffic, noise, odours, smoke, dust, litter and pests.

From bad to worse...

European laws are forcing the Government to send less waste to landfill and the landfill tax is rising to deter businesses and local authorities from landfilling waste which could be recycled. The money raised from landfill tax should be used to provide every home in the UK with a doorstep collection for recycling and composting.

However, the Government is instead:

■ Replacing landfill with incinerators.
■ Dumping toxic incineration ash in existing landfill sites.

Friends of the Earth says:

■ Don't build any more incinerators.
■ Use money from landfill taxes for recycling and waste reduction.

■ The above information is reprinted with kind permission from Friends of the Earth. Please visit www.foe.co.uk or see page 41.

© *Friends of the Earth*

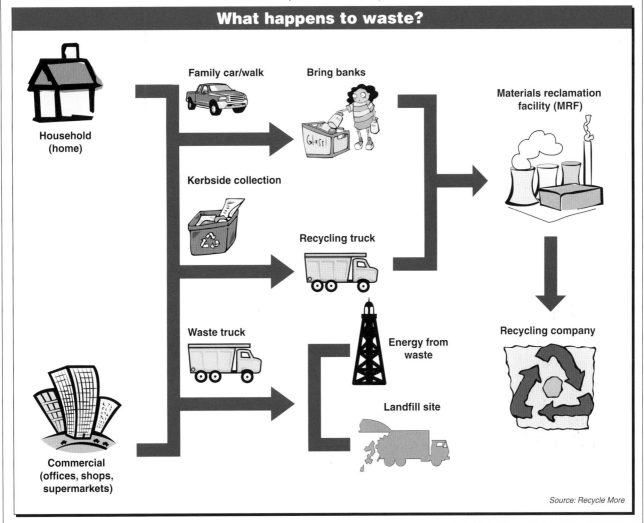

What happens to waste?

Family car/walk

Bring banks

Materials reclamation facility (MRF)

Household (home)

Kerbside collection

Recycling truck

Waste truck

Energy from waste

Recycling company

Landfill site

Commercial (offices, shops, supermarkets)

Source: Recycle More

Quick facts about litter

Information from EnCams

The role of Keep Britain Tidy

ENCAMS (Environmental Campaigns) runs Keep Britain Tidy, a campaign aimed at getting the public to dispose of materials appropriately, so they do not become litter. ENCAMS also provides free publications and information on litter-related issues to the public, through a helpline number (01942 612639) and its website (www.encams.org).

Litter measurements

ENCAMS carries out surveys across the country measuring different sorts of litter, including: smoking-related litter, fast food packaging, confectionery litter and hazardous litter such as broken glass and syringes. The information is used to detect trends in litter deposits, to help local authorities to make more efficient use of their resources and to inform ENCAMS' future litter campaigns.

Litter campaigns

ENCAMS has conducted a number of campaigns over the years. A recent example was a youth campaign aimed at 13- to 16-year-olds. All secondary schools in the UK were sent a set of six posters using aspirational figures

from *Coronation Street*, *Brookside* and Liverpool Football Club (including Michael Owen). Pop groups SClub and Atomic Kitten were also used with slogans supporting 'Keep Britain Tidy'. Schools have reported a very positive response. Further posters were given away at cinemas, with ENCAMS being inundated with requests for more, and running out within a week of the campaign starting. The posters were also featured on 3,000 Adshel (bus stop) sites across the country for two weeks over the Christmas period, in addition to a small number of 48 sheet poster sites.

ENCAMS conducts public campaigns such as this on a regular basis. For current campaign details visit the ENCAMS website at www.encams.org.

Cost of clearing up litter

It can cost a council between £6 and £19 per kilometre to sweep the streets, depending on the nature of the area and the methods used. The direct cost of street sweeping and cleaning was £413 million in 1999-2000. The cost for 1998-1999 was £400 million.

Public policing

Even though it is infuriating to see someone littering, ENCAMS does not advocate putting personal safety at risk. The same applies if you see litter thrown from cars. Police officers or litter wardens are empowered and trained to deal with offenders. If you have information about a littering incident you could report it to the police, the local authority or a litter warden, and it is up to them to decide whether they wish to proceed any further. Whilst it is possible to take a private prosecution, it would be at your own expense and you will need strong evidence to prove your case in court.

Litter clearance

The local authority has a legal duty (so far as practicable) to clear litter and refuse from public places for which it is responsible, such as streets, parks, playgrounds, tourist beaches and pedestrianised areas. If a piece of private land is littered, the owner should accept responsibility for clearing the litter. Often, however, the problem is determining who owns the land. Each local authority will have its own title for the department that carries out street cleansing. Usually it will be called something like Cleansing, Environmental Services or Environmental Health. You can phone them and report litter problems.

Schools and litter

Schools have a legal duty to clear the litter and refuse from their own grounds. This is laid down in the Code of Practice on Litter and

Refuse. If they fail to do so, the council can take out a Litter Abatement Notice against them (served on the Chairman of the Board of Governors) to force a clean-up – it is an offence not to comply. Also, you can personally take out a Litter Abatement Order against a school to get them to clean up. However, the school has no responsibility for clearing the litter outside its grounds. It is the council's duty to clean streets and other public places. Children over 13 can be prosecuted for littering, but this rarely happens.

Lack of bins
For security reasons, a number of bins have been removed in the last decade, particularly in railway stations and on the Underground. Bins have to have a particular specification for Health and Safety reasons and cost between £400 and £800 each. They must be waterproof, UV resistant and fire resistant. The siting of a bin can require planning permission that can take up to two years. Each bin has to be emptied, cleaned and maintained.

Litter prosecutions
In 1990 there were 2,543 litter prosecutions. In 1999 there were 501 litter prosecutions and, of those, nearly 90% of those prosecuted were male.

Legal definition of litter
Legally speaking, the word litter is given a wide interpretation. Litter can be as small as a sweet wrapper, as large as a bag of rubbish, or it can mean lots of items scattered about. ENCAMS describes litter as 'Waste in the wrong place caused by human agency'. In other words, it is only people that make litter. The offence of 'Leaving Litter' (section 87 of the Environmental Protection Act 1990; Article 3 of the Litter (NI) Order 1994) says that if a person drops, throws, deposits or leaves anything so as to cause defacement in a public place, they could be committing a littering offence.

Fly-tipping
This is a word to describe the act of dumping rubbish illegally. Often this can be seen outside waste amenity sites, and includes larger items such

as furniture or business waste. It is, of course, illegal and, being potentially more serious and hazardous than littering, is an offence under waste legislation. On conviction, the fine is up to £20,000 or six months' imprisonment or both.

Most prevalent litter
From the ENCAMS Local Environmental Quality Survey of England completed in early 2002, the most commonly found types of litter across all locations and land use are:
- Smoking-related litter (77% of all locations)
- Confectionery-related litter (53% of all locations)
- Drinks-related litter (31% of all locations).

Smoking-related litter
In research carried out by ENCAMS (1996), 96% of smokers aged 21–25 admitted that they threw their cigarette ends on the ground, and 46% of all smokers claimed they had never used a bin for their cigarette ends. The main reason quoted was the perceived fire hazard.

Litter perpetrators
From the 2002 survey of England (see above), ENCAMS found the worst perpetrators to be pedestrians or drivers. They cause 87% of litter. Only 3% is from domestic bins (being torn or spilling over), and only 4% is commercial/business waste.

Public disgust at littering
Most of the population voice their disgust at seeing litter in the streets. However, research by ENCAMS (October 2001) shows that nearly

every adult in the population drops litter in varying degrees. The most frequently littered items were those that were seen as small and less harmful, such as small sweet papers, apple cores, cigarette butts and chewing gum. Eighty-seven per cent of litter is caused by the public, and all sections of society are contributing. However, most people do not consider that they are adding to the problem, but blame the following instead:

96% of smokers aged 21-25 admitted that they threw their cigarette ends on the ground, and 46% of all smokers claimed they had never used a bin for their cigarette ends

---

- Teenagers
- Education in schools
- The council
- Not enough bins
- Bins not emptied frequently enough
- Too much packaging and wrappings
- Not enough prosecutions
- Fines not high enough.

Public confessions!
Research by ENCAMS (October 2001) shows that the public admit the most likely situation in which they would drop litter is when they are driving. They feel that they are non-accountable in a car and can't be identified. They also think it is

more acceptable to drop litter if an area is already run-down or dirty and if there aren't sufficient bins. They would think twice about littering if they were in their own neighbourhood, if the area was tidy and presentable or if they were with children.

Litter in Singapore

It is often mentioned that Singapore does not have a litter problem. However, Singapore has solved its litter problem by adopting stringent enforcement that we would probably find unacceptable in the UK. Aside from general laws on litter, there is also a prohibition on the importing of chewing gum. The penalty for littering is a fine up to $10,000 or 12 months in prison or both. Second offences carry a penalty of $20,000 fine and/or two years in prison. Rumour has it that the ban was imposed to keep the subway running on time after used wads of gum, disposed of on train doors, prevented them from closing and disrupted services. Litter offenders are counselled to stop them reoffending, and corrective work orders are imposed as a form of community service (10 hours' cleaning a housing estate for leaving a paper clip on the branch of a tree, for example). The courts are not afraid to impose high fines for serious offences on a 'tiered' basis, for example a drinks can is considered worse than a cigarette end.

The public admit the most likely situation in which they would drop litter is when they are driving

Litter longevity

Some litter can last longer than others, often taking years to degrade. Plastic bottles can last indefinitely, and plastic dropped in a field today could still be there in the next millennium. Aluminium litter such as drinks cans last from 80 to 100 years, as do nappies. Plastic bags given away free by supermarkets last between 10 and 20 years. Cigarette butts, orange peel, banana skins and apple cores can last up to two years.

■ The above information is taken from the EnCams document *Litter and the law – a guide for the public*. Visit www.encams.org.uk to view the entire document or for more information, or see page 41 for contact details.
© EnCams

Packaging waste

Information from the Chartered Institution of Wastes Management

Packaging waste includes glass, paper, board, metals and wood. It makes up over half of our household waste and is growing in volume and weight. Because of the growing importance of packaging in the waste stream it was the first subject of 'Producer Responsibility' controls by the European Union.

The Packaging Waste Directive (94/62) came into force in 1994 with clear objectives:
■ reducing over-packaging;
■ eliminating dangerous materials from packaging;
■ reducing the proportion of packaging waste going to landfill;
■ increasing recovery and recycling of packaging waste;
■ putting the burden for recovery and recycling on the 'producer';
■ providing consumers with information.

Under producer responsibility the 'producer' is obliged to ensure that target proportions of the materials put into the market are recycled or have value recovered from them.

In the UK Producer Responsibility Obligations (packaging waste) Regulations 1997 the 'producer responsibility' is shared between 4 different sectors:
■ converting raw material into packaging materials, e.g. oils into plastics, 6% of the total responsibility.
■ converting materials into packaging items, e.g. plastics into plastic bottles or containers, 9%.
■ packing or filling products into containers – e.g. fizzy drinks into plastic bottles, 37%.
■ wholesalers / retailers, 48%.

About 5,000 companies in England and Wales are affected by these controls. They must register with the appropriate agency, supply a certificate of compliance and either recover / recycle the waste packaging themselves or register with a 'compliance scheme' who will do it for them. If your annual turnover is more than £2 million, or if you handle more than 50 tonnes of packaging annually then you need to contact your environment agency for information.

■ The above information is reprinted with kind permission from the Chartered Institution of Wastes Management. Visit www.ciwm.co.uk for more information or see page 41 for address details.
© CIWM

Food packaging

**Boxes, bottles, packets, cartons and cans –
information from TheSite.org**

The global food packaging industry is now worth $100bn a year, growing 10-15% each year. Anything between 10% and 50% of the price of food today can be down to its packaging.

How much waste?

3.2m tonnes of the 26m tonnes of the household waste produced annually are packaging. 150m tonnes of packaging waste come from industry and commerce each year. To achieve a change towards more sustainable packaging, it's not just the packaging that requires alterations but also our lifestyles and habits of consumption.

People in the US chuck away 2.5m plastic bottles every hour and less than 3% are recycled. But we aren't so squeaky clean in the UK: of the 15m plastic bottles used every day, much less than 3% gets recycled. Fewer than 1% of the billions of plastic bags we use each year are recycled and the majority are used only once. European law wants us to recover 50% of all our packaging and to recycle 25%, but Britain, predictably, is lagging seriously behind.

TheSite.org

Why use packaging?

- Packaging provides a physical barrier between a product and the external environment thereby ensuring hygiene and reducing the risk of product wastage due to contamination.
- Some forms of packaging prolong the life of food.
- Some packaging is also needed for safe and efficient transportation.
- Packaging is also used to provide customers with information and instructions, for which there are some legal requirements.

Why so much waste?

- The decrease in the size of households over the decades has resulted in more people buying smaller portions of food, and thus more packaging.
- Higher living standards in the western world have led to more consumer goods and to the transportation of exotic foods over long distances requiring a large amount of packaging to maintain freshness.
- A trend towards urbanisation, which creates longer distances between food producers in rural areas and consumers in urban areas, has also led to a demand for packaging.
- Other contributing factors are the increases in working families (i.e. both partners working) along with the increase in fridge and freezer ownership, which has led to a higher demand for convenience food.

Has it all gone too far?

Frankly, yes. Food packaging today is really about marketing, in an increasingly consumer world, food producers care more about competing for shelf-space than environmental consequences, and we, the consumers, aren't much better. Do you consider the impact arising from the toxins and pollutants released at every stage in production and transport? Do you think about how long the packaging will take to biodegrade or incinerate, or about how some products are just economically unviable to recycle in rich countries and instead are shipped out to pollute the third world? Probably not, but maybe you should.

Packaging materials

The most common types of material used for packaging are paper, fibreboard, plastic, glass, steel and aluminium.

- **Paper:** one of the most widely used packaging materials, particularly corrugated cardboard used for transport packaging. The current recycling rate for paper and board packaging waste is 49%.
- **Glass:** is the most common form of packaging waste. Glass can be returned and reused or recycled easily and a well-established recovery and recycling system exists in the UK. The first bottle-bank in the UK appeared in 1977 and today there are over 20,000. Six billion glass containers are used annually in the UK and the recycling rate is 22%.
- **Aluminium:** is used in many packaging applications such as beverage cans, foils and laminates. It has a high value as a scrap metal with prices ranging from 650-750 pounds per tonne and can be recycled economically. Of the estimated 5 billion aluminium cans consumed in the UK in 1996, 31% were recycled.

- **Steel:** is a widely used packaging material for food, paint and beverage as well as aerosols. Recycling steel brings significant resource and energy savings. The current recycling rate for steel cans is 16%.
- **Plastic:** offers several advantages over other packaging materials in its sturdiness and low weight. Even though plastic can be recycled there is a lack of facilities in the UK. The current recycling rate for plastic in the UK is 5%, with the remainder either land-filled or incinerated.
- **Mixed materials:** packaging can sometimes have the benefits of being more resource and energy efficient than single material packaging, but combining materials makes recycling difficult. Recycling these materials is hindered by the lack of facilities and technology necessary to separate materials to avoid contamination. Mixed materials packaging can be reprocessed into other products such as floor coverings, shoe soles and car mats, incinerated to produce energy, or landfilled.

- The above information is reprinted with kind permission from TheSite.org. Please visit the website at www.thesite.org for more information.

© TheSite.org

Pigeon population plump and rising

Information from EnCams

Bird-brained louts who dump fast food rubbish on the streets are not only making our environment look unhealthy – they're also creating a race of super-sized pigeons, dependent on people rather than nature for their food.

And, argues charity Keep Britain Tidy, unless these junk food addicts shake their dirty habits and use a bin, the pigeon population will get so out of hand that councils may be forced to cull THOUSANDS of our fattened feathered friends.

'Seven out of the ten bits of litter we find on our pavements and roads are food related,' said Alan Woods, Chief Executive of the campaign. 'And with all this nosh to choose from, the pigeon, rat, fox and gull

population has spiralled. This isn't fair on the councils who are left to control this pest problem and is cruel to those animals who are scavenging in unnatural environments for food that isn't good for them.'

Alan's message is particularly poignant at this time of year – when the hot weather encourages people to eat al fresco. June and July is also peak breeding time for pigeons. To try and curb our whopping food litter problem, Keep Britain Tidy is backing a Government initiative called, 'Food on the Go' which encourages everyone from sandwich shops to drive-thrus to keep their premises clean and push the anti-rubbish message to their punters. They also welcome new legislation in the Clean Neighbourhoods and Environment Act, making it easier to hit litterers with £50 on-the-spot fines.

But Keep Britain Tidy isn't only pointing the finger at litter louts for creating a plump pigeon population. They are also backing a drive launched in Derby by the Cleaner Greener Normanton Project, to discourage people from feeding urban birds. This campaign will include visits to community centres and places of worship and will see flyers pushed through doors to ensure the message hits home.

Fast food rubbish - fat lot of bad

- The amount of fast food rubbish found on England's streets has risen by 50% in the last three years!
- An increase in the number of drive-thrus has meant that some fast food trash is being found three to four miles away from where it was bought.
- Takeaway debris doesn't only litter our pavements, the grease and fat from it also stains them (grime is a particular problem at shopping parades and precincts).
- But it isn't only food that gets dropped. Dumped drinks cans are found on around half of our streets.
- According to vets, 69,000 animals per year are injured by litter. One of the most common complaints they treat is dogs and cats who have choked on chicken bones and apple cores.
- Apart from swelling the pigeon population, dumped food has also helped hike rat numbers by 24% in the last three years.

'People genuinely feed pigeons out of a sense of kindness,' continued Alan Woods. 'But by leaving food around, they are not helping the birds at all. Pigeons become dependent on you for their diet and when flocks gather, this spreads disease and drives smaller birds away. Really, the best way to care for pigeons is to stop dropping and littering food – and instead let nature takes its course.'

Coo-ee – fascinating facts about flappers

- Pigeons have only 37 taste buds which means that though their natural food is grain, meat or green vegetable matter, they will practically feed off anything.

- Human food does not contain the nutrients necessary to a pigeon's good health.
- In some parts of Frankfurt, Germany, animal lovers have set up pigeon hotels to house and feed their feathered friends. This is an effort to stop the city from culling them!
- Before London's mayor Ken Livingstone prevented the sale of seed at Trafalgar Square, 300,000 pigeons flocked there, daily. The bill for cleaning up their poop stood at £100,000 a year.
- The average life-span of a pigeon is three to five years – but some have lasted as long as 35 years.

- To try and minimise the damage droppings do, councils and other landowners pigeon-proof their buildings to stop birds from roosting there. Other more bizarre solutions have included feeding pigeons contraceptive pills and placing plastic hawks on buildings – to frighten them away.
- Litter hasn't only made its way into pigeon's bellies, nests have also been found that contain junked bits of plastic.

18 July 2005

- Information from EnCams. Visit www.encams.org for more information or see page 41 for contact details.
  © EnCams

Study serves up nappy dilemma for parents

Information from the Environment Agency

Parents will need to do more than just think about the type of nappy they buy if they want to look after the environment according to a major study published today (19 May 2005) by the Environment Agency.

The study, which looks at and evaluates the environmental impacts arising from every stage of the life cycle of disposable and reusable nappies, found that there was little or nothing to choose between them.

Tricia Henton, Director of Environmental Protection at the Environment Agency, said: 'This study was carried out to establish the true environmental impacts of using disposable and reusable nappies. Although there is no substantial difference between the environmental impacts of the three systems studied, it does show where each system can be improved.

'We hope manufacturers of disposable nappies will use this study to improve the environmental performance of their products, particularly the quantities going to landfill. Similarly, if parents using reusables want to improve their impact on the environment they will need to look more closely at how they launder nappies, for instance, can the nappies be washed in a bigger load at a lower temperature?

'The type of nappy that parents buy is a matter of personal choice, but it is important that they can make an informed choice. Studies like this help to inform people about the impact that their actions have on the environment. Life cycle thinking plays an important role in informing the environmental debate.'

The study found there was little or nothing to choose between disposable and reusable nappies

---

For each nappy type studied, all the materials, chemicals and energy consumed during manufacture, use and disposal were identified and the resulting emissions to the environment accounted for.

The study is based upon an average UK child wearing nappies for the first two and half years and is supported by behavioural surveys carried out in 2002/3 covering more than 2,000 parents and guardians of children who use or have used nappies.

It looked closely at a wide range of activities associated with disposable and reusable nappies which affect the environment. These included:

- the energy and material used in the manufacture of the nappy;
- the daily number of changes required for the different types of nappy;
- how reusable nappies are washed – temperature, size of loads, type of detergents etc.

The study, which was carried out by independent environmental consultants, is the most comprehensive and thorough independent study of its kind ever undertaken in the UK.

19 May 2005

- The above information is reprinted with kind permission from the Environment Agency. Visit www.environment-agency.gov.uk or see page 41 for more details.
  © Environment Agency

The nappy debate

Environment Agency nappy report is seriously flawed

A long and expensive study comparing the environmental impacts of disposable and cloth nappies is seriously flawed, says Women's Environmental Network (WEN).

The lifecycle analysis (LCA) for the Environment Agency has taken four years and cost more than £200,000 and concludes there is 'no significant difference between any of the environmental impacts'. But the conclusion is based on poor quality data and misses the point of its own findings.

'This lifecycle analysis is a wasted opportunity to put the long-standing debate about nappies and the environment to rest,' says WEN's Ann Link. 'It says what most other LCAs have: that both systems use similar amounts of energy but the disposable system uses more materials and puts more into landfill. But it has missed the point of its own findings. Even in its current flawed state it shows that parents who use cloth nappies can save waste confident in the knowledge that washing them will cause no more global warming than disposable nappies.

Women's Environmental Network

'The biggest impacts it identifies are all to do with energy production and use – abiotic resource depletion (fossil fuel use), global warming and acidification – yet if parents use 24 nappies and follow manufacturers' instructions to wash at 60 °C using an A-rated washing machine they will have approximately 24% less impact on global warming than the report says.'

The LCA measures the environmental impact of reusable and disposable nappies from the raw material stage (e.g. coal, trees or cotton plants) through manufacturing processes to their use, disposal and emissions (e.g. CO_2) back into the environment.

Data used for key assumptions are unsatisfactory in many instances. The report itself raises concerns with regard to assumptions that result in 'a high level of uncertainty associated with the reusable nappy systems'. It admits that 'the amount of analysis and quality of the results might be improved with a larger sample and by refining the questions'.

Over 2,000 parents using disposable nappies were surveyed. By contrast, most of the survey results for reusable nappies were drawn from a sample of 117 parents, further reduced to 32 because users of terry towelling nappies were relied on for most assumptions. This resulted in as few as two respondents being used for certain key assumptions.

The study gives little useful advice and will confuse parents.

The reusable system has enormous potential for improvement in environmental performance. This is hinted at in the LCA but not highlighted. Consumer guidance from sensitivity analyses has not been provided on what practice would achieve a significant reduction in environmental impact. WEN has found that a 17% reduction in global warming impacts can be achieved by using an A-rated washing machine and following manufacturers' guidance to wash at 60 °C. With A-rated washing machine sales at near saturation by early 2005 many real nappy users are already achieving this saving. Parents only need use 24 real nappies, rather than the 47 the LCA assumes, reducing their global warming impact by another 6.9%.

Notes

Waste: Britain throws away nearly eight million nappies a day. With a disposal cost to individual local authorities in hundreds of thousands of pounds per year (Nottinghamshire estimates £1 million per year) it is not surprising that nappy schemes now play a key role in local authorities' waste strategies.

Cost: WEN estimates that washing nappies at home could save parents around £500. Hospitals can save money too by using real nappies on wards where disposables incur clinical waste charges. Local authorities save on waste disposal charges.

Modern nappies have advanced considerably over recent years. They are shaped and fitted and fastened without the need for pins. They come in a variety of styles and patterns. A biodegradable liner can be used inside the nappy and this can be removed so that the contents are flushed down the loo. Nappy washing services make things even easier, collecting dirty nappies and leaving fresh clean ones in their place.

While WEN welcomes the Environment Agency's undertaking to carry out further work in this area it is urgently required to bring the report in line with recent technological and product developments.

Disposable nappies are the largest single product category in household waste. While waste amounts are still rising, the EU landfill directive requires a reduction of 35% in biodegradable waste over a 25-year period.

Elizabeth Hartigan, co-ordinator of WEN's real nappy project, says: 'Using real nappies puts parents in control. With a good washing routine parents can minimise the environmental impact of their babies' nappies, reduce waste and save themselves money.'
19 May 2005

■ Information from the Women's Environmental Network. Visit www.wen.org.uk or see page 41.
© WEN

The rubbish problem

Frequent questions

1. Why is rubbish a problem?
The amount of waste produced in this country is a clear sign that our economy is environmentally unsustainable. 81% of our rubbish ends up in landfill, which means we are literally using up and throwing away the earth's natural resources. Many of these resources are renewed by nature over time but if we continue to use them up faster than they are replenished, eventually there'll be none left.

88% of household rubbish is disposed of in landfill sites. As the waste decomposes, it produces dangerous substances

---

Many plastics, for example, need quantities of oil to produce them – what will happen when the oil runs out? Living within the limits of the earth's natural systems will mean using less and being more efficient.

2. What's so bad about burying our rubbish?
88% of household rubbish is disposed of in landfill sites. As the waste decomposes, it produces two dangerous substances:
1. Methane gas, which contributes to the greenhouse effect causing climate change.
2. Leachate, which can seep from older landfills into our water supply causing pollution.

There are 2,300 landfill sites in the UK. Existing landfills are predicted to be full within 5-10 years. Rubbish from cities is often transported many miles by large lorries burning lots of fuel and creating more pollution.

3. So why not burn more of our rubbish?
Sending waste up into thin air sounds like the perfect solution for getting rid of it and the energy that comes from sending waste up in smoke can provide us with electricity.

Unfortunately, incineration can cause problems of its own. Burning waste can release a range of pollutants into the atmosphere including heavy metals, gases that cause acid rain and gases which contribute to climate change. The ash that is left over is often toxic and has to be disposed of in ... yes, you've guessed it – landfill sites. Incinerators need lots of rubbish to run economicially, meaning recycling can be sqeezed out of local markets.

There are currently incinerators in the following places:
SELCHP, South East London; Edmonton, North London; Bellingham, Cleveland; Bolton; Tysely, Birmingham; Stoke; Coventry; Nottingham; Sheffield; Wolverhampton; Dudley; Isle of Wight; Dundee; Shetland; Kirklees, near Huddersfield.

Even more on the way at: *Chineham, near Basingstoke; Capel, Surrey; Bexley; Portsmouth; Southampton; Allington, Kent; Ridham Dock, Kent; Richborough, Kent; Newhaven; Robertsbridge, East Sussex; Derby; Grimsby; Wrexham; Swansea; Aberdeen; County Antrim; Kidderminster; Hull; Guildford; Redhill, Surrey; Horsham; Basildon; Colchester.*

4. Is recycling the answer?
It's part of it, along with cutting down on the rubbish we create. Recycling means making something new from something old – for example, drinking glasses from recycled bottles. Although the recycling process uses energy and water it usually isn't as much as making a product from scratch. Recycling also cuts down on raw materials having to be extracted from the earth's resources.

5. I'm already recycling. What more should I do?
If you are already recycling, there are still ways you can reduce your rubbish. Think about reducing the overall amount of rubbish you produce each week, or about what happens to rubbish you produce at work.

■ Reprinted with permission from Global Action Plan. For more information please visit their website at www.globalactionplan.org.uk or see page 41 for their address details.
© Global Action Plan

Waste prevention and you

Information from the Women's Environmental Network

Every year, we use more and create more waste. Recycling is still important, but to really make a difference, we need to reduce waste and reuse things. Do you want to cut back on the amount of waste you produce? Not sure how?

Here are some top waste prevention tips:

1. Think about what's in your rubbish – is any of it reusable? Repairable? Do you really need to throw it out?

In Britain, we each generate an average of 522 kg of waste per year. That's the same as 10 bags of sugar a week

---

2. Make a waste prevention plan – Take a look around your home and at what you buy and use the checklists to make your own waste prevention plan. If you start off with the small things, preventing waste will soon become second nature.

Waste prevention checklists
At home
- Avoid single-use disposable products – e.g. nappies, tissues, facewipes, razors, polystyrene and plastic cups, plates and cutlery, kitchen towels, serviettes, computer cartridges, cameras.
- Buy refills of cleaning products and toiletries.
- Avoid buying lots of cleaning products.
- Return junk mail and remove your name from junk mail lists www.mpsonline.org.uk.
- Repair and mend items rather than throw them away.
- Avoid battery-powered toys or use rechargeable batteries.

- Try washable menstrual products (check out the Sanpro section of the WEN website).
- Use a milk-delivery service or organic box scheme with refillable containers.

In the kitchen
- Don't make too much food and keep your leftovers for the next day.
- Buy in bulk and store food in reusable containers instead of foil or clingfilm.
- Make food at home instead of buying takeaways or fast food to avoid packaging waste. If you do buy takeaways, ask your local takeaway to use more environmentally-friendly packaging.
- Use towelling face washers or cotton/linen napkins instead of paper ones.
- Drink tap water and reuse water bottles.

At the shops
- Only buy what you need – 2 for 1 offers sound great, but do you really want or need more than one?
- Take your own shopping bag (WEN sells lovely reusable cloth bags) and say NO to plastic ones.
- Shop locally – walk, cycle or use public transport.
- Buy locally produced goods whenever possible.
- Avoid over-packaged products and try to buy unpackaged goods.
- Support repair shops.
- Buy experiences instead of things (e.g. trips to the theatre or a massage).
- Remove excess packaging and leave it in the shop along with a word of protest.
- Buy products made from recycled materials.
- Donate furniture, computers and white goods and clothes to reuse projects and buy second-hand.

In the office
- Print/photocopy on both sides, proofread and spellcheck your work before printing.
- Reuse envelopes by using sticky labels – buy packets of recycled labels from WEN for £3.20.
- Treat yourself to a refillable ink pen.
- Consider a wormery for all those office tea bags!
- Buy recycled paper.
- Recycle your toner cartridges.
- Write and put in place a green housekeeping policy.
- Encourage your staff to cycle to work.

In the garden
- Ask your council to supply home composters or wormeries or take a look at the Waste Links section of the WEN website to see how to get hold of one.
- Compost your kitchen and garden waste.

Campaigning
- Start up or join a local waste prevention or composting group (see the WEN website).
- Ask your local authority what they are doing about waste prevention (see WEN's *25 Ideas For Local Authorities* on their website).
- Contact other organisations with an interest in waste prevention.
- Distribute awareness-raising materials. WEN has a full range of resources for you to use and adapt.

Quick fact: In Britain, we each generate an average of 522 kg of waste per year. That's the same as 10 bags of sugar a week!

- Information from the Women's Environmental Network. Visit www.wen.org.uk for more information or see page 41 for contact details.

© WEN

What can you do about waste?

Information from Recycle More

There are a number of ways to reduce and recycle rubbish in the home, and at school. In most cases it is better to choose items which create less rubbish, for example goods without excessive packaging.

Use our waste diary to record your rubbish [see the Recycle More website].

Look at the areas below for more ideas of how you can help cut down on the amount of rubbish you produce. You could reduce the rubbish in your bin by over 50%.

Reduce

- don't buy heavily packed goods.
- buy 'loose' food rather than pre-packaged.
- stop junk mail and faxes through the Mailing Preference Service.
- cancel delivery of unwanted newspapers, donate old magazines to waiting rooms.
- use your own shopping bags when visiting the supermarket or use the doorstep delivery service.
- grow your own vegetables. Many varieties can be grown in small gardens.
- use a nappy laundry service, and save disposable ones for holidays and long journeys.
- take a packed lunch to work or school in a reusable plastic container.

Reuse

- reuse carrier bags. Each person in the UK uses an average of 134 plastic bags each year.
- reuse scrap paper for writing notes, etc.
- reuse envelopes – stick labels over the address.
- rent or borrow items you don't use very often – e.g. party decorations and crockery. Some supermarkets hire out glasses for parties, saving on disposable cups.
- donate old computer and audio visual equipment to community groups or schools.
- buy rechargeable items instead of disposable ones e.g. batteries and cameras.
- buy things in refillable containers e.g. washing powders.
- buy concentrated products which use less packaging.
- take old clothes and books to charity shops, or have a car boot sale.

- look for long-lasting (and energy-efficient) appliances when buying new electrical items – ensure these are well maintained to increase product life cycle.

Use your own shopping bags when visiting the supermarket

Recycle

- choose products in packaging which you know can be recycled.
- compost – lots of kitchen waste can be composted. Contact your local council for details of local composting schemes and details of any compost bin sales.
- buy products made from recycled materials. Most supermarkets now stock a wide range of these items.
- find out where your nearest recycling facilities are.

■ The above information is reprinted with kind permission from Recycle More. Please visit their website at www.recycle-more.co.uk for more information or if you wish to write to them, please see page 41 for their contact details.

© Recycle More

Waste disposal methods

BEST ENVIRONMENTAL OPTION → **LOWEST**

REDUCE — Reducing rubbish means trying to stop it being created in the first place. For example, reducing the amount of packaging around products.

REUSE — What's rubbish to you, may be used again by someone else. A good example of this is a glass milk bottle that your milkman will collect and reuse.

RECOVERY — When people talk about the 'recovery' of rubbish, they mean that the rubbish has been recycled, composted or burnt (incinerated) to produce heat or electricity. Recycling means the rubbish has been turned into something useful.

DISPOSAL — Usually it's best to try and reuse or recover the rubbish, but if we can't then we can bury the rubbish in a rubbish tip (also called a landfill site).

Source: Recycle More

Waste in the workplace

Information from Waste Watch

All businesses produce waste, whether this is solid waste for recycling or disposal, emissions to air such as gases, or waste water. Arranging for the disposal of these wastes and complying with legal requirements can be costly and time-consuming. But the true cost of waste is much more than the cost of its management and disposal.

There is a 'hidden' cost to the waste businesses produce. This includes the loss of raw materials (which have been purchased in the first place, but which end up as waste), as well as the time and energy invested in processing the raw materials. It has been estimated that UK businesses lose up to 4.5% of annual turnover every year through avoidable waste.

Reducing your waste is about being efficient. Put simply, reducing your waste means that you use less (and therefore spend less), get more out of what is used, and reuse or recycle any unavoidable wastes. By increasing efficiency in this way, businesses can maximise their outputs and increase profits, whilst also saving valuable resources and helping the environment.

So saving resources and cutting down on waste not only makes business sense, but also helps keep our environment clean and safe!

Businesses are recognising the important role they have to play in helping the UK to achieve more sustainable patterns of production and consumption. Depending on the business and the activities undertaken there are a variety of ways in which products can be manufactured, or goods/services provided, in better ways, using fewer resources. Some of these are simple to introduce and will have immediate benefits, such as making better use of paper in the office or implementing water-saving measures. Others may require more effort but are likely to bring significant benefits in the longer term, such as changing a process so that fewer resources are used or re-designing a product so that it is easier to repair or dismantle.

There are many ways in which businesses – producers and retailers alike – can become waste-wise. Here are some simple first steps.

- identify areas for waste reduction and implement a waste reduction strategy
- support reuse, take-back and refurbishment schemes
- identify the non-hazardous wastes produced by your company which can be recycled and set up a recycling scheme
- encourage employees to reduce, reuse and recycle at work and at home
- keep abreast of – and be compliant with – forthcoming legislation.

For further information and advice on how to reduce your waste, contact envirowise, Wastebusters or the Environment Agency.

- The above information is reprinted with kind permission from Waste Watch. For more information visit www.wastewatch.org.uk or see page 41 for address details.

© Waste Watch

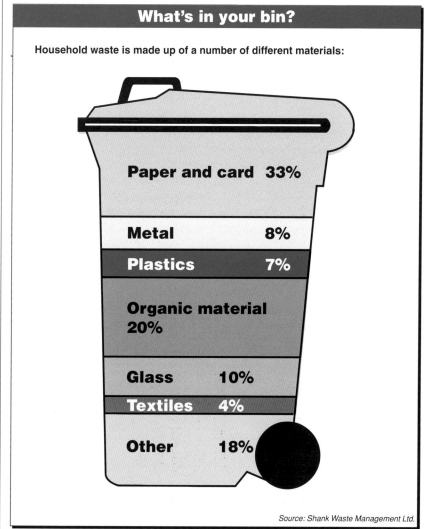

What's in your bin?

Household waste is made up of a number of different materials:

- Paper and card 33%
- Metal 8%
- Plastics 7%
- Organic material 20%
- Glass 10%
- Textiles 4%
- Other 18%

Source: Shank Waste Management Ltd.

Litter louts face new on-the-spot fines

By Ben Willis

Dropping litter even in your own back garden will now be considered a criminal offence under one of three last-minute bills passed by parliament before it dissolves next week.

Local authorities and parish councils will be given new powers to impose fines and clearance notices on individuals and businesses to remove litter from their land.

It is estimated that the annual burden of litter, abandoned vehicles and noise on the public purse is £919m

---

The new measures form part of the Cleaner Neighbourhoods and Environment Act, which was given royal assent last night (7 April 2005) along with the Disability Discrimination Act and the Drug Act.

The Cleaner Neighbourhoods Act forms part of the government's attempt to improve cleanliness and safety in the urban environment. It is estimated that the annual burden of litter, abandoned vehicles and noise on the public purse is £919m.

The act significantly increases the powers of local authorities to operate without police assistance in imposing fines for a range of offences including littering, antisocial behaviour, flytipping and abandoned cars.

It is understood that on-the-spot fines for individual littering could be around £75, and as much as £200 for abandoned cars, although councils will have the flexibility to set their own local rates.

Welcoming the new powers, David Sparks, chairman of the Local Government Association's environment board, said: 'A rundown neighbourhood is more than just an eyesore – it can increase the fear of crime, hamper economic regeneration and result in a loss of local pride. This new act will enable quicker and more effective enforcement that will deter offenders.'

Local authorities will also be given new powers to 'alley-gate' – close off public rights of way, usually behind houses, that are magnets for antisocial behaviour.

This measure has caused concern among campaigners, who claim that it could lead to restrictions on freedom of movement.

Hester Brown, parliamentary officer of lobby group Living Streets, said: 'The act gives councils the authority and incentive to take action that will lead to positive results, but we would have liked to have seen the alley-gating measures strike more of a balance between the need to protect residents from antisocial behaviour and the right of people to walk freely around their neighbourhood.'
8 April 2005

Fly tipping

Information from the Chartered Institution of Wastes Management

Fly tipping is the illegal deposit of any waste onto land. This includes the 'tipping or dumping' of waste onto land that does not have an appropriate licence to accept waste. Some fly tipping is low volume but with a big impact on local environmental quality: white goods and furniture in verges, builder's rubble in gateways, plastic sacks in alleys. However, incidences can be more serious involving more hazardous wastes such as clinical waste, tyres or asbestos, or drums of liquid. The circumstances and locations can be many and various – at the least fly tipping is unsightly and costly for landowners to remove. At its worst it can be polluting and / or dangerous to human health. Fly tipping also avoids the cost of proper waste management – undercutting more responsible businesses.

One of the most important controls helping to combat fly tipping is the Duty of Care supported by Waste Carrier Registration.

Anyone finding or witnessing fly tipping should contact their environment agency or local authority.

Fly tipping is an offence under the Environmental Protection Act 1990, the Refuse Disposal (Amenity) Act 1978 and various Highways Acts. The penalty for infringement is a heavy fine and can lead to a custodial sentence and loss of vehicle if applicable.

Because of the nature of fly-tipped wastes and its occurrence a number of authorities and agencies can become involved. To overcome any potential duplication and to ensure a comprehensive approach, the Fly Tipping Stakeholders Forum has produced guidance for landowners, managers and members of the public.

■ The above information is reprinted with kind permission from the Chartered Institution of Wastes Management. Visit www.ciwm.co.uk for more information or see page 41 for their address details.

Dumping on Britain

Dealing with illegal dumping

Recent figures show that waste is illegally dumped in England every 35 seconds and costs £100 a minute to clear. Much of this is dealt with by your local council. The Environment Agency deals with the big, bad and nasty stuff that can cause problems for human health and the environment. The dumping of asbestos is something that we take very seriously. The dumping of asbestos only accounts for 8.3 per cent of the incidents we deal with.

We deal with around 5,000 incidents of fly-tipping every year.

The kind of changes we are seeing are:

- Increase in big, bad and nasty waste crime such as fly-tipping of construction waste which can be very hazardous to the environment.
- Increase in illegal exports and imports of wastes.
- Illegal dumping gangs operating across a wider geographical area.
- Evidence of more organised and 'career criminals' committing environmental offences.

What kinds of waste crimes are committed?

Waste regulations are in place to protect the environment and human health so we take waste crime very seriously. Examples of the sort of crimes we've seen are:

- Businesses operating several illegal sites and transfer stations and failing to dispose of waste in the correct way. One operator in Wales was estimated to have earned about £1 million from this kind of activity in the last two to three years.
- Illegal importation of biodegradable waste for disposal, covertly disguised as non-hazardous waste for recycling and recovery.
- Gangs accessing vacant industrial premises, changing locks and filling the sites with wastes having charged unknowing businesses to take the waste away to be disposed of properly.
- In the North East, one waste criminal received £80,000 over six months to remove tyres. None were properly disposed of.
- Organised criminals with links to terrorism in Northern Ireland laundering red diesel caused a water pollution incident and abandoned hazardous waste residue. We also think that the same gangs may have been involved in illegal drug manufacturing which leaves hazardous residue behind.

Why is this kind of crime on the up?

There are a number of factors which have caused this increase in serious waste crime:

- The costs of complying with waste management regulations are increasing. So it's tempting to cut costs and dump waste illegally.
- The financial rewards can be lucrative.
- Where the police and others are working to tackle serious crimes such as drug crime, criminals will turn to something they see as less risky – like environmental crimes.
- The availability of suitable waste sites can make it difficult for legitimate waste management companies.

And what are we going to do about it?

We are working more closely with our partners like the police and local authorities to detect offenders. For example, we've set up an online service – Flycapture, which will help us gather intelligence about serious offenders.

We're also providing our enforcement officers with more training and helping them develop new skills as cases become more complex. We're modernising our approach to regulation and aim to implement a more joined-up approach across all of the regions where we regulate.

- Information from the Environment Agency. For more visit www.environment-agency.gov.uk

© Environment Agency

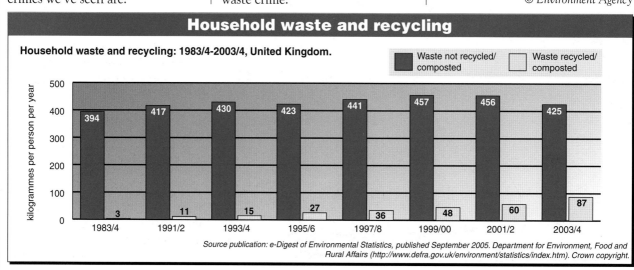

Household waste and recycling

Household waste and recycling: 1983/4-2003/4, United Kingdom.

Legend: ■ Waste not recycled/composted □ Waste recycled/composted

Year	Waste not recycled/composted	Waste recycled/composted
1983/4	394	3
1991/2	417	11
1993/4	430	15
1995/6	423	27
1997/8	441	36
1999/00	457	48
2001/2	456	60
2003/4	425	87

(Y-axis: kilogrammes per person per year)

Source publication: e-Digest of Environmental Statistics, published September 2005. Department for Environment, Food and Rural Affairs (http://www.defra.gov.uk/environment/statistics/index.htm). Crown copyright.

The Clean Neighbourhoods and Environment Act

Fly-tippers to pay for environmental crime

Fly-tipping is now costing almost £1 million a week to clear up, according to data released by Defra today (18 Oct 2005).

However, from today the Clean Neighbourhoods and Environment Act gives local authorities and the Environment Agency new powers to recover these costs from the offenders.

The Clean Neighbourhoods and Environment Act

The Clean Neighbourhoods and Environment Bill received Royal Assent and became the Clean Neighbourhoods and Environment Act on 7 April 2005. In June 2005 Defra announced the first raft of the Act's measures, which included fines of up to £50,000 and five years in prison for those found guilty of fly-tipping.

Today the second raft of the Act's measures have come into force. Other new measures which come in to force today mean:

Department for Environment Food and Rural Affairs

- landowners and occupiers who have had to clear fly-tipped waste can also recover costs.
- local authorities have more effective investigatory powers to catch fly-tippers.
- local authorities have the power to remove abandoned cars from the streets immediately rather than giving 24 hours' notice.

Flycapture

The latest data on fly-tipping have been gathered from Flycapture, the national database developed by Defra and the Environment Agency. Most local authorities have been submitting data to Flycapture for the past 12 months. It contains information on incidents dealt with, actions taken and an offenders' vehicle registration hotlist. Graphs and data are available on this site: www.defra.gov.uk/environment/localenv/flytipping.

£44 million of council taxpayers' money

Announcing the latest measures, alongside the 2004/05 Flycapture data, Margaret Beckett said:

'Around one million incidents of fly-tipping were recorded on Flycapture last year, costing local authorities more than £44 million to clear up. That's £44 million of council taxpayers' money that could be spent improving other council services.

Around one million incidents of fly-tipping were recorded on Flycapture last year, costing local authorities more than £44m to clear up

---

'But from today, offenders themselves will foot the bill: under the Clean Neighbourhoods and Environment Act, local authorities and the Environment Agency will now be able to make the polluters pay for the mess they have made, as well as the costs of tracking down the culprits.'
18 October 2005

- Information from Defra. Visit www.defra.gov.uk for more information or see page 41 for their address details.

© Crown copyright

Recycling

Information from the Young People's Trust for the Environment

We produce more rubbish today than ever before, on average each household in Britain produces about a tonne of waste every year. Much of this waste contains potentially useful materials such as paper and board, glass, metals and textiles which could be recycled, reducing the amount of rubbish, creating less pollution and saving energy.

Rubbish

The amount of rubbish we produce has been escalating over the last 40 years. There has been a gradual change in shopping habits and people's attitudes to throwing things away. The personal service provided by shopkeepers has been replaced by self-service in supermarkets where the goods are often highly packaged; often loose items are packed together and priced to speed up payment at the checkouts. Some goods are elaborately wrapped to make them look more attractive, put into plastic bags and then loaded into plastic carrier bags at the checkout. A Women's Environmental Network group bought a trolley-load of 102 basic items – the shopping for a family for two weeks. They found that there was a total of 543 pieces of packaging with some items wrapped in up to five layers!

On average a family of four throws away about two sacks of rubbish a week, most of which could be recycled. The figures below show the main constituents of household waste:

- Paper and card 18%
- Garden waste 21%
- Kitchen waste 17%
- Glass 7%
- Metal and white goods 8%
- Plastic 7%

(*Waste and Resources Action Programme (WRAP) 2002 c/o www.wasteonline.org.uk*)

The amounts are quite staggering. Each person in a year generates 10 times their own weight in household rubbish, throwing away an estimated 90 drink cans, 107 bottles and jars, two trees' worth of paper, 70 food cans and 45kg of plastic.

Landfill

Almost 90% of domestic waste in the UK goes directly to landfill or dumping sites to be levelled and covered with earth. This costs about £1 million a day. Once the rubbish has been covered the organic matter starts to rot down producing methane, an inflammable gas, which makes its way to the surface. In some tips the methane is piped off and used as fuel for heating. Landfill waste remains a potential environmental hazard. Weedkillers in the rubbish, chemicals from car batteries and other dangerous liquids can be washed through the soil, contaminating drinking water. In landfill sites where toxic industrial wastes have been dumped indiscriminately, the land can become poisoned and unsafe for farming or building. Today, waste disposal is regulated by a number of European Community directives which help to ensure that the disposal of waste is controlled and safe.

Incineration

Ten per cent of all domestic waste is burned in waste incinerators but this method of disposal can be hazardous. If the temperature in the incinerator is allowed to fall below 900°C, some plastics, pesticides and wood preservatives can produce dioxins which are extremely poisonous.

ACTION

Each pupil could make a survey of what is thrown away at home using bar graphs to keep records. This will give some idea of the enormous amount of resources which are thrown away each week.

Packaging

Packaging materials make up about 7% of household waste. Traditional materials like waxed card and paper have been replaced by foamed plastics, aluminium and polythene. The result is a high volume of lightweight rubbish which swamps litter bins, blows about in the wind and is, in the main, non biodegradable.

Packaging has three main purposes: protection, preservation, and to make the product look more attractive to potential buyers. Many packaging materials are combined together in such a way that they are impossible to separate and therefore cannot be recycled.

ACTION

If there is no choice, buy products which are contained in the least amount of packaging. Buying in bulk helps since items packaged in small quantities produce more waste than those packaged in large quantities. Packaging made of paper or cardboard is preferable to plastic and glass bottles are better than plastic ones, especially if they are returnable.

The problem with plastic

More than 3.5 million tonnes of plastics are used in the UK each year, making up a further 7% of household rubbish. Nearly all the plastics in use are made from oil and resist any form of biological decomposition. These are non-biodegradable plastics and cause problems in waste incineration since many of them give off poisonous gases when burned. Biodegradable plastics which are made from sugar and other carbohydrates rot away within months of being buried. However, the cost of biodegradable plastics is far greater than that of ordinary plastics since their production is carried out on a small scale. If biodegradable plastics were widely used in preference to other plastics manufacturing costs would drop dramatically.

Recycling organic waste

Organic waste includes anything which is biodegradable such as food waste, newspaper, eggshells, natural fibre clothes, wood shavings, leaves and other garden waste. The more water organic matter contains, the faster it will decay. Compost for the garden can be made from organic matter which is broken down by earthworms, bacteria and other decomposers releasing nutrients which help plants to grow.

Recycling glass

Glass makes up another 7% of household rubbish. Every year, the average family uses about 500 glass bottles and jars in Britain, the equivalent of nine items a week per household. Glass is completely recyclable yet we throw away 1.5 million tonnes of it a year. Every tonne of old glass used saves 135 litres of fuel and replaces 12 tonnes of raw materials. Despite these obvious advantages only about 15% of the glass used in Britain is recycled compared with over 50% in Europe as a whole. By December 2006, we need to reach a recycling rate of 60% to meet the EU Packaging Directive.

Returnable bottles

Even more energy is saved if bottles are not melted down but are reused instead. A collection system already exists for milk bottles which are returned for reuse up to 20 times. Bottles are the easiest containers to reuse and many manufacturers could be using them instead of plastic or metal. Although bottle banks encourage the use of throwaway bottles, in the absence of bottle reuse schemes they are better than nothing.

ACTION

Set up a project to examine how many different designs of bottles would be needed if all bottles were returned for reuse. Aspects to be taken into consideration would be the viscosity of contents, accessibility, width of neck for pouring, the effect on contents of clear or dark glass, strength of glass, stability, packaging, sealing etc.

Recycling metals

Iron, steel, tin, copper and aluminium can all be recycled. The recovery of iron and steel has been carried out in the UK for more than a century and is still an important source of raw materials for steel making.

Cans

In 2000, the UK consumed 5 billion aluminium can drinks, of which about 42% were recycled. Cans for food and drink are made from aluminium or tin-plated steel. Any kind of can may be recycled. De-tinning salvages the tin lining which protects steel cans from rusting; aluminium and de-tinned steel can both be smelted for reuse. Steel cans can be extracted from household waste with a magnet. The cans then go directly to the steel plant for recycling.

The main problem in can recycling is the quantity of cans needed to make a scheme viable. A collection scheme is operated by the Can-Makers' Association and skips are usually placed on sites near supermarkets or in car parks.

Recycling aluminium

Aluminium is a relatively new but rapidly increasing element of household waste. About 75% of the drinks cans we use are made entirely from aluminium. There is a constant demand for them since they can be recycled again and again.

Aluminium is one of the most expensive and potentially most polluting metals to produce. It is extracted from bauxite ore mined at the surface. The open-cast mines cover large areas from which the natural vegetation has to be removed. To extract the metal requires huge quantities of electricity, much of it coming from hydroelectric power stations. Dams are built across valleys and large areas inundated by the lakes that form behind them. While this form of power does not create the air pollution problems associated with electricity generated by fossil fuels, the construction work and lakes upset the local ecology. In the process of extracting the metal, fluorides can be emitted into the atmosphere. These damage the health of workers and plants and animals near the smelter. Aware of the damage that can be done to the environment, the industry has gone to great lengths to reduce its impact. The land at the

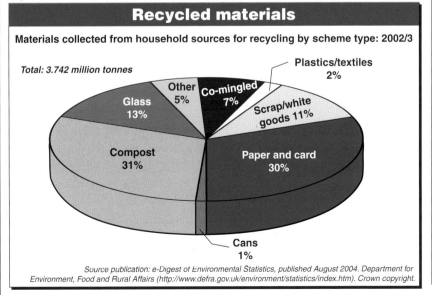

Recycled materials

Materials collected from household sources for recycling by scheme type: 2002/3

Total: 3.742 million tonnes

- Other 5%
- Co-mingled 7%
- Plastics/textiles 2%
- Scrap/white goods 11%
- Glass 13%
- Compost 31%
- Paper and card 30%
- Cans 1%

Source publication: e-Digest of Environmental Statistics, published August 2004. Department for Environment, Food and Rural Affairs (http://www.defra.gov.uk/environment/statistics/index.htm). Crown copyright.

mine is restored and re-vegetated once it is no longer needed. The amount of energy needed to smelt a tonne of ore has been reduced by 40% over the past 40 years and the amount of fluoride getting into the atmosphere has been reduced to very low levels.

Melting down an aluminium can for reuse requires just 5% of the energy needed to make a new can and it creates little pollution. The quality of the recycled aluminium is just as good as the primary metal, so cans can be recycled again and again. Every can that is thrown away is a lost opportunity to save energy and preserve the environment.

ACTION

Charities often raise funds by collecting and recycling crushed cans. Some cans have a symbol on the side identifying them as aluminium. If not, use a magnet to test the side of the can (some steel cans have an aluminium top and bottom). If it is not magnetic then it is made of aluminium. Put the ring pull inside the can and then crush the can and store it for collection.

Recyling paper

At present less than 25% of the paper in use is recycled and the rest is produced from wood pulp. Increasing the volume of recycled paper could reduce the pressure on the world's valuable timber resources.

Recycled paper has to be de-inked and then chemically treated to separate the fibres. Different grades of waste paper have different recycling values: waste paper from offices which is high-quality paper and only lightly inked, is particularly valuable.

The manufacture of recycled paper uses only half the amount of energy and water used in making virgin paper. New paper is often white, not because this is paper's natural colour but because it is bleached. The bleach used in making white pulp can cause appalling water pollution. Toxic wastes such as dioxin are amongst the other wastes discharged from pulp mills. Yet until recently there was little control of the waste being discharged. Recycled paper tends to be slightly grainy and grey or light green. There is no need to use bright-

---

The manufacture of recycled paper uses only half the amount of energy and water used in making virgin paper

---

white paper when a high quality recycled paper could easily be substituted for most uses.

Waste paper can also be used as animal bedding, fuel logs and pellets and, suitably treated, as insulation in the home. Waste products can become a resource with a little ingenuity.

ACTION

Making recycled paper

Soak some old newspapers in a bucket overnight and drain off the extra water. Using a liquidiser or wooden spoon mash the paper and water into a pulp. Put the pulp into a bowl, add an equal volume of water and mix these together. Slide a wire mesh into the mixture and lift it out covered in pulp. Lay a cloth on a clean, flat surface. Place the mesh, with the pulp side down, quickly and carefully onto the cloth. Press it down hard, then peel off, leaving the pulp on the cloth. Put another cloth on top and press down firmly. Repeat these steps with the remaining pulp. Place a plastic bag on the top and weight the pile down. After several hours gently peel the paper off the cloths. Leave the pieces on some kitchen towel until completely dry. The paper should now be ready for use.

Recycling oil

Many people are still unaware that oil can be recycled. In 1999, 380,000 tonnes of oil were disposed of, sometimes illegally poured down drains where it can be a serious pollution problem, instead of returning it to garages or local authority collection points. Even a few litres of oil spilled into a lake can produce a thin film over thousands of square metres and kill plants and animals. A refining company can produce up to 3,500 litres of usable oil from every 5,000 litres of processed waste oil. It can be used as a lubricating oil or further processed for use as a heating oil.

What can be done

- Don't mix up your rubbish. Separate glass, paper, metal and organic waste.
- Buy and use recycled paper products wherever possible. A list of stockists is available from Friends of the Earth.
- Use your own shopping bag. Do not accept plastic carrier bags every time you shop.
- Refuse to buy over-packaged goods. Buy fresh unpackaged food whenever possible. Look out for goods in recyclable containers.
- Contact your local authority and ask about their waste disposal and recycling policies.
- Contact Oxfam and other charities to find out what they are collecting. Items collected for sale might include aluminium foil, cans, clothing and newspapers.

Industrial waste

The safe disposal of hazardous industrial waste has become a very high priority for governments and industry as new products and materials come into wider use. Disposing of waste such as plastic, solvents, oils and heavy metals poses problems that did not exist or were not understood a few decades ago. Many dangerous substances dumped in the past are still causing environmental problems today.

The disposal of hazardous waste is now controlled but many conservationists are still concerned that the standards set are too low to protect the environment adequately. They also claim there are too few inspectors to make sure that the proper procedures are adhered to, especially at sea.

- Information from the Young People's Trust for the Environment. For more information visit www.yptenc.org.uk or see page 41 for their address details.

Going green

Recycling reaches all-time high

British Glass
Manufacturers' Confederation

'We are turning into a nation of "recycling champions",' claimed Olympic hero Sir Matthew Pinsent as he launched the BIG recycle today (24 June 2005). The four-time Olympic gold medallist revealed that UK households and businesses are recycling more waste packaging than ever before, according to a major new study.

The new figures have been released to coincide with 'the BIG recycle' – a week-long recycling campaign that begins on 27 June. They reveal:

- UK households recycled around a third of all their packaging in 2004 – an increase from around a quarter in 2002.
- Approximately 3.5 billion glass bottles and jars, 1 billion plastic bottles, 2 billion aluminium cans and 2.5 billion steel cans were recycled by UK households in 2004. This is a total weight, including cardboard containers, of 1,253,000 tonnes of material.
- Recycling in UK business (offices and industry) is also up. In 2004, industry used 5.6 million tonnes of packaging – 66 per cent of which was recycled. This is an increase of seven per cent on 2002.
- Almost half of all the packaging used in the UK is recycled.
- Recycling is getting more convenient. Almost every council in the UK now provides some form of service for doorstep collection of recyclable or compostable materials.
- Nine out of 10 people regard recycling as an important thing to do, and 50 per cent are classed as committed recyclers.

The new figures are from Pack-Flow, a seven-month project led by producer responsibility scheme Valpak with the involvement of materials organisations British Glass, Corus, Alupro and Recoup, and WRAP (the Waste & Resources Action Programme). The project surveyed local authorities across England, Wales, Scotland and Northern Ireland, to get a comprehensive national picture of packaging recycling rates.

Now in its second year, 'the BIG recycle' is part of Recycle Now, a major TV and press advertising campaign designed to encourage people in England to recycle 'more stuff, more often'. It will see events taking place across England, Scotland and Wales, organised by local authorities, retailers and community groups to encourage the public to recycle more household waste.

Four-times Olympic champion Sir Matthew Pinsent, and Sarah Beeny, presenter of Channel 4's *Property Ladder*, are both backing the campaign. In addition, Paralympic champion Tanni Grey-Thompson and former Olympic athlete Roger Black are supporting the BIG recycle.

Matthew Pinsent said: 'I've always been a big believer in recycling and use my local council's doorstep collection service. I don't even have to think about it now – I just do it. If we all realised what a massive difference we could make by re-

glass jars & bottles

cycling used packaging, such as food and drinks cans, glass and plastic bottles and jars, I think we would all do it.'

'The BIG recycle' is jointly organised and funded by WRAP – a UK programme established to promote resource efficiency – in partnership with the leading materials recycling organisations British Glass, Corus, Novelis, PaperChain, the Corrugated Packaging Industry, Recoup and Valpak.

Ben Bradshaw, Parliamentary Secretary for the Department for the Environment, Food and Rural Affairs, said: 'I welcome this new study which shows that households and businesses are recycling more than ever before, which is excellent news. There is no doubt that it is becoming easier to recycle in the UK, and that people are becoming increasingly keen to do so.

'However, there is still room for improvement. The study does indicate that meeting European 2008 packaging waste targets will be challenging. We need to continue expanding our recycling infrastructure, and encouraging all households and businesses to play their part. Awareness-raising campaigns such as "the BIG recycle" have an important role in shifting public attitudes.'

Jennie Price, Chief Executive Officer of WRAP, said: 'Recycling is one of the easiest ways for people to reduce their personal environmental impact, so this week would be a great time to recycle just one extra bottle, can or newspaper.

'The BIG recycle is all about letting people know what, where and how to recycle so they can see for themselves just how easy it is to recycle more stuff more often.'

- The above information is reprinted with kind permission from British Glass. For more information visit www.britglass.org.uk

Recyclable materials

Information from Shanks Waste Management Ltd

Aluminium

Aluminium is a type of metal which is used a lot to make drink cans because it is light and unbreakable.

There would be 12 million fewer dustbins per annum in the UK if all aluminium drinks cans were recycled

---

Aluminium can also be rolled into very thin sheets and used to make foil, takeaway containers, pie cases and milk bottle tops.

Recycling tips
- Check whether a type of packaging is aluminium by looking for the symbol.
- Rinse and crush cans by standing on them before you recycle them.
- Aluminium foil cannot be recycled with aluminium cans. Check with your local council for aluminium foil recycling schemes in your area.

Amazing facts
- Recycling an aluminium can saves up to 95% of the energy needed to make a new one.
- Recycling 120 aluminium cans saves enough energy to run a 3-bedroom house for one day.
- If all of the aluminium cans recycled in the UK were laid end to end they would stretch from Land's End to John o'Groats 160 times!
- There would be 12 million fewer dustbins per annum in the UK if all aluminium drinks cans were recycled.

Steel

Steel has become a very common type of packaging used by manufacturers because it is very strong and lightweight.

Think about all the packaging you buy in the supermarket which is made of steel, including:

- All types of food cans.
- Pet food tins.
- Some drinks cans.
- Aerosols, toiletries and paint containers.
- Bottle tops and jam jar lids.

Recycling tips
Before you put your steel cans out to be recycled:
- It is easy to check whether packaging is steel by using a magnet, as steel will stick to it!
- Rinse them.
- Put the lids inside them.
- Put any jam jar lids or bottle tops inside the cans.
- Squash the end of each can flat, so that it takes up less space.

Amazing facts
- 3 out of 4 of the cans in the supermarket are made of steel.
- All new steel products already contain 25% recycled material.
- In the UK we use 13 billion steel cans each year. If we stacked them on top of each other they would reach the moon 3 times!
- Steel cans are also called 'tins' because they are coated with a very fine layer of tin – another metal.

Plastics

Plastic is used excessively by packaging manufacturers today because it is lightweight but is also very strong, so it won't break easily.

These qualities mean that it is a perfect material for packaging juice, milk, toiletries, cleaning products and fizzy drinks into plastic bottles.

The 3 main types of plastic bottles which can all be recycled are:
- HDPE.
- PET.
- PVC.

Recycling tips
Before you put your plastic bottles out to be recycled:
- Take off the lids.
- Rinse the bottles.
- Squash the bottles, so they take up less space.

Amazing facts
- In the UK, 23 million plastic bottles are produced every day.

- Recycled plastic bottles can be recycled into a range of products including pipes, building products and even clothing!
- It only takes about 25 recycled soft drink bottles to make one fleece jacket.
- Recycling just one plastic bottle can save the same amount of energy needed to power a 60-watt light bulb for 6 hours.

Paper

Paper is made by using wood chips so recycling paper saves trees, which means less forests need to be cut down.

Think about all the paper you use every day – writing paper, envelopes, printer paper, newspapers, magazines, junk mail, leaflets, directories, wrapping paper…

Recycling tips
- Don't tie newspapers and magazines together with string as it isn't recyclable!
- Try to keep paper dry where possible.

Amazing facts
- Reclaimed waste paper represents around 63% of the fibre used to produce paper and board in the UK.
- On average every person in the UK gets through 38kg of newspapers a year.
- The average household throws away 3 kg (7lbs) of paper every week! That works out to be the average family throwing away six trees' worth of paper each year.
- Every year we need a forest the size of Wales to provide all the paper we use in the UK.
- 17 trees are needed for one tonne of paper, that's about 7,000 copies of a national newspaper.
- Paper can be recycled up to 6 times.

Glass

Glass is used to make all sorts of bottles and jars for packaging food and drinks.

Glass bottles and jars come in 3 main colours, clear, brown or green.

Glass is 100% recyclable. It takes less energy to melt recycled glass than to melt new raw materials, so recycling saves both natural resources and energy.

Waste management by region

Management of municipal waste in 2003/04 by region, England.

Notes: Totals from returned questionnaires with estimates for missing authorities. Totals might not add up due to rounding.

Source publication: Municipal Waste Management Survey, published July 2005. Defra. Crown copyright.

Recycling tips
- Use the bottle banks for recycling, located at supermarkets, car parks and household waste sites.
- Separate glass into the different colour grades, brown, clear or green. They all contain slightly different ingredients and must be recycled separately.
- Remove any bottle caps, lids or corks from glass bottles and jars.
- Rinse glass jars out to remove any food.
- Never put pottery, crockery or heat-resistant glass (such as Pyrex) in a bottle bank.

Amazing facts
- Glass packaging makes up about 9% by weight of all the rubbish we throw away.
- There are 22,000 bottle-bank sites in the UK.
- Every glass bottle recycled saves enough energy to light a 15-watt energy-efficient light bulb for 24 hours.
- Almost 200 glass jars and bottles are thrown away in the UK every second.
- In the UK, we use over 6 billion glass containers each year. It would take 3,500 years to sing '6 billion green bottles hanging on a wall'.

Cardboard

Cardboard is mainly used as packaging to make boxes and other containers because it is so strong and lightweight.

Cardboard is really just another type of paper which is much thicker and therefore stronger.

It is just as important to recycle cardboard because, like paper, it is made from wood pulp, produced from cutting down trees.

Recycling tips
- Squash boxes flat before you take them for recycling.
- Don't worry about removing staples or tape from the boxes.
- Don't leave plastic wrapping inside boxes.

Amazing facts
- Paper and cardboard makes up 32% (1/3) of all the household rubbish that we throw away.
- We use 9 million tonnes of paper and cardboard every year in the UK.

Paper can be recycled up to 6 times

- Cardboard consists of two different types of paper, smooth paper called liner and crinkle paper called fluting. Cardboard is made by putting three layers together, a liner on the top and bottom and the fluting in the middle.
- Some recycled cardboard is so strong that you can use it to make things that might have otherwise been made from wood or metal, like furniture and large containers.

- Information reprinted with kind permission from Shanks Waste Management Ltd. Please visit www.shanks.co.uk for more.

© Shanks Waste Management Ltd

Is your brain full of rubbish?

All right, let's see how much you really know about waste and the 3Rs!

Try these eight questions to test your knowledge on waste and what should be done with it.

1. What is the best way to deal with our rubbish?
a) Reuse it
b) Recycle it
c) Reduce it
d) Throw it in the bin

2. What is the best way to reduce the number of bottles and cartons wasted at school?
a) Don't drink anything at school
b) Use bottles that can be filled up again and again
c) Use bottles that can be recycled
d) Use smaller bottles and cartons

3. Which of the following helps to reduce the amount of paper used?
a) Putting it in a recycling bin
b) Putting it in a dustbin
c) Writing or printing on both sides
d) Making papier mache models

4. What do the nappies that an average child uses in his or her lifetime weigh the same as?
a) A bicycle
b) A family car
c) A lorry
d) A motorbike

5. What fraction of our household waste could be recycled?
a) More than a half
b) Less than a quarter
c) Nearly all of it
d) About a third

6. If you collected up all of the empty aluminium drinks cans in the UK and took them for recycling, how much money could you make?
a) 2 thousand pounds
b) 30 million pounds
c) 17 thousand pounds
d) 1 million pounds

Answers: c, b, c, b, a, b.

Scores 0 to 2
Your brain is nowhere near full enough of rubbish.

Scores 3 to 4
Good but there's still plenty more room for rubbish in your brain.

Scores 5 to 6
Congratulations. Your brain is even more full of rubbish than your bin!

Now that you know so much, make sure that you put your knowledge into practice by remembering to reduce, reuse and recycle whenever possible.

© *Waste Watch (www.wastewatch.org.uk)*

Five reasons to recycle glass

Information from British Glass

1. Glass recycling saves energy
Making new glass from recycled glass uses much less energy and reduces CO_2 emissions.

The energy saving from recycling one bottle will:
- Power a 100-watt light bulb for almost an hour.
- Power a computer for 25 minutes.
- Power a colour TV for 20 minutes.
- Power a washing machine for 10 minutes.

Every household in the UK uses on average 331 bottles and jars per year. If the average household recycled all their glass they would save enough energy to:
- Power a 100-watt light bulb for 12.5 days.
- Power a computer for 5 days.
- Power a colour TV for nearly 4.5 days, enough to watch 210 episodes of *Coronation Street*.
- Power a washing machine for 2.5 days.

2. Glass recycling conserves the environment
Recycling your glass saves raw materials from being quarried and then thrown away in rubbish dumps as used bottles and jars. This saves hundreds of thousands of tonnes of quarrying each year and conserves the countryside for everyone.

3. Glass recycling creates employment
Jobs are created by glass collection schemes and at recycling centres, which crush and clean the recycled glass.

4. Glass recycling increases public awareness of the problem of rubbish
Everyone can help the environment by recycling their glass, even a small change in behaviour has a measurable benefit. This is a first step towards becoming environmentally active.

5. Glass recycling cuts waste disposal costs
By weight, glass makes up about 8 per cent of our rubbish. Glass recycling reduces the cost of collecting and disposing of glass mixed in with our rubbish.

- The above information is reprinted with kind permission from British Glass. For more information visit www.britglass.org.uk

© *British Glass*

Uses of recycled waste

Information from RecycleNow

Products and packaging made wholly or in part from recycled materials are gradually becoming more widespread and can usually be identified by the inclusion of the mobius loop and a percentage upon their packaging. Although they are becoming increasingly frequent, recycled products and packaging remain the victim of unjustified beliefs that recycled products are in some way inferior to comparable items made from new. In many people's eyes, 'made from recycled materials' has become synonymous with poor quality, expense, unattractiveness and inefficiency. In reality though products made from recycled materials can be indistinguishable from their counterparts manufactured from virgin materials.

Aluminium, steel and glass can all be recycled repeatedly without a loss in quality. Some recycled paper is now manufactured to exceptionally high standards and will even function equivalently in printers and photocopiers. There are few reasons why recycled products should cost any more than new and you should find that most will actually be competitively priced. It is not necessarily true that recycled products are unappealing to look at either. For some products it is impossible to tell the difference between recycled and new and for many others,

innovative design that hints at the item's original use forms part of the appeal.

For more information on products made from recycled materials and a recycled products guide listing over 1,000 items, please visit www.recycledproducts.org.uk

■ The above information is reprinted with kind permission from RecycleNow. For more information visit www.recyclenow.co.uk

© Budget Pack

Doorstep recycling survey

Information from Friends of the Earth

Friends of the Earth completed a survey last year which found that almost three-quarters of councils in England, Wales and Northern Ireland were failing to provide a basic recycling service to all of their residents.

The survey showed that there is still a long way to go before local authorities meet the minimum standards set out in the Household Waste Recycling Act.

Key findings from the research, completed in March 2004

■ Only 114 local authorities (29 per cent) comply with the Household Waste Recycling Act's 2010 target, offering all their households a doorstep collection of at least two materials.

■ On average local authorities serve 65 per cent of households with a doorstep collection of two or more materials.

■ 'Best practice' doorstep collections are more scarce. Less than half of local authorities provide a collection for five or more materials. Only 26 per cent of households on average are served with this type of collection.

■ But the news isn't all bad as doorstep recycling services are growing, with 92 per cent of local authorities now offering a doorstep collection for two or more materials to some households.

Best practice doorstep recycling

Local authorities are being urged to follow Friends of the Earth's 'best practice' code to improve the collection services offered to householders and increase participation in recycling schemes. Measures include:

■ Collecting a wide range of materials on a weekly basis.
■ Providing better information to householders.
■ Increased effort to reach out to 'difficult' properties such as high-rise and rural dwellings.

'Some local authorities run excellent recycling schemes. These need to be copied across the country if England is to have a recycling record to be proud of' – Claire Wilton, Waste Campaigner.

■ The above information is reprinted with kind permission from Friends of the Earth. Visit www.foe.co.uk for more information or see page 41.

© Friends of the Earth

Should I bother recycling?

Leo Hickman's guide to a good life

Blame the Minoans. They had the bright idea to dig large pits away from their homes, fill them with their rubbish, then cover the lot over with earth. Three thousand years on, and not much in municipal waste disposal has changed, except that where they left something to excite archaeologists by throwing out their broken amphoras, we now mark our place in history with discarded tyres, plastic bags and empty cereal packets.

We are finally starting to confront our 'out of sight, out of mind' attitude towards our rubbish

It's taken a bit of time, but we are finally, albeit with the speed of a turning tanker, starting to confront our 'out of sight, out of mind' attitude towards our rubbish. Or rather, we have been forced to confront it as available landfill space rapidly runs out or faces death by EU directive. The options now before us are stark: we can consume less and thus throw out less, we can try to reuse or recycle as much of our rubbish as possible, or we can wallow in our putrescent effluence and be damned.

Sadly, given our nation's addiction problem when it comes to shopping, the first option seems a distant fantasy; a bit of a downer when you consider that 80% of what we buy ends up being discarded within six months and that UK homes are currently tossing out 26m tonnes a year (40m tonnes by 2020). Which leaves us – assuming we don't want to face the last option – with recycling. Cheeringly, whether it's because of a growing awakening to the problem by us, the tossers, or through increasingly persistent arm-twisting by local councils (themselves under pressure due to the threat of missed targets and fines), we are starting to adopt the habit of recycling. In fact, last week the government even gave us a little pat on the back for our efforts. The amount of household waste being recycled over the past four years has doubled, it said.

The sound of party poppers, however, is dampened somewhat when you learn how our recycling and composting rates – on average, 17% – compare with other countries. Swiss households, for example, recycle 53% of their waste.

But is recycling really the answer? What of the oft-heard claims that it can take more energy to sort, collect and recycle something than it does to produce it in the first place? And what of the cynical belief that what we send to be recycled is just tipped, out of view, on to landfill anyway?

Having once visited my local recycling centre, in large part to put my mind at rest about the latter point, I can confirm that what we put out for recycling is definitely separated and sorted. I stubbornly stood and watched my bottles – with about 40 tonnes of other bottles – being driven off in a truck to be reprocessed. The energy-comparison argument comes down on the side of the recyclers, too. According to Waste Watch, the

energy used to recycle plastic bottles is eight times less than required to manufacture the same virgin polymer. Producing recycled paper uses up to 70% less energy than virgin paper, as well as using far less water. And recycling just one glass bottle saves enough energy to power a TV for 20 minutes.

There are some caveats, though. 'Trash miles', like food miles, are becoming a problem. Much of what we recycle is sent abroad, often to China where worker conditions can be poor, because there is simply no market here for most of the materials. For example, half the plastic bottles we diligently put out for recycling are now thought to be sent to China where they command a price of £50 a tonne, whereas here they are virtually worthless. China has quickly become the world's leading rag-and-bone man, but what some see as a sensible market-driven solution can seem short-sighted when you factor in transport-related emissions and the health of the workers (who often sort through toxic waste, such as old computers, with their bare hands).

It would be much better – and laws are soon to start enforcing this – if much of what we bought was designed with reuse or recycling in mind. It would be better still if we interrupted the waste stream by curbing our overall consumption, as well as buying goods made from recycled materials where possible (Recyclenow.com has examples). Recycling should really be seen as the last resort.

20 September 2005

You say . . .

'I'm committed to recycling, but until the supermarkets learn to stop over-packaging goods that local councils are not prepared to recycle, we are fighting a losing battle. I return all unnecessary packaging to my supermarket. If everybody did this, it would send a message that consumers are tired of paying for unwanted packaging' – Susanna White, Bath.

'Recycling would be worth it if I didn't have to drive out of town to do it, only to find the paper bank full and a sign telling me to chuck it in the household waste bins' – Claire Lackford, by email

---

Recycle

Information from Global Action Plan

Recycling is on the increase in the UK, but there's still a long way to go. This article offers some tips, but recycling facilities will vary from area to area.

To find out more about recycling facilties in your own area visit www.reuze.co.uk, www.Recycle More.co.uk or contact your local authority.

Paper and cardboard

Recycling paper reduces pressure on natural resources and uses 30-70% less energy than producing paper from virgin materials. There should be a paper bank near you – make sure you deposit the right type of paper in the right recycling bank.

Metals

Simple – wash and recycle all your tins and cans. Don't forget your aerosol cans are recyclable, just like any other steel or aluminium container.

Plastics

If you're not one of the lucky few to live near the 2,851 plastic bottle collection points in the UK then recycling your plastic can be difficult. Visit www.recoup.org for more details on different types of plastic, where you can recycle them and how to buy recycled plastic products.

Electronic equipment

Action Aid's National Recycling Scheme for printer cartridges and mobile phones. Oxfam shops will take back your old mobiles, as should the retailer. Over 1.5 million computers are dumped in landfill every year in the UK. Throwing other electric equipment like fridges, televisions, or cookers into landfill is environmentally damaging, but unfortunately there are few take-back facilities for these products. This will change, when the EU WEEE Directive comes into force in 2005, and local authorities will have to provide a free service for households.

Batteries

Batteries are pretty nasty beasts. Many older types of battery contain potentially harmful metals like mercury and cadmium, and there are so many different sizes and types of batteries that sorting and recycling is difficult. Rechargeable nickel cadmium batteries, like those in drills or mobile phones, can be recycled: www.rebat.com. Other types of battery are more tricky to recycle – the RABITT scheme offers some recycling services . The best plan? Use batteries as little as possible.

Glass

Glass is one of the best materials for recycling, as it can be recycled again and again, saving energy and raw materials. Before recycling your glass, make sure you wash the bottles and jars, and remove any tops or plastic attachments. Do try and put your glass in the right coloured banks – any contamination will lower the value of the recycled glass.

Garden and kitchen waste

Organic waste is the main cause of methane emissions, a powerful greenhouse gas, from landfill sites. Instead of binning it, try composting. It's an easy, satisfying process; not only will your dustbin be less smelly, but you'll be improving your local environment and even saving money into the bargain. If you don't have a garden to compost in, some local authorities have set up community composting sites. Visit the Community Composting Network or the Composting Association.

■ The above information is reprinted with kind permission from Global Action Plan. Please visit www.globalactionplan.org.uk for more information or see page 41.

© Global Action Plan

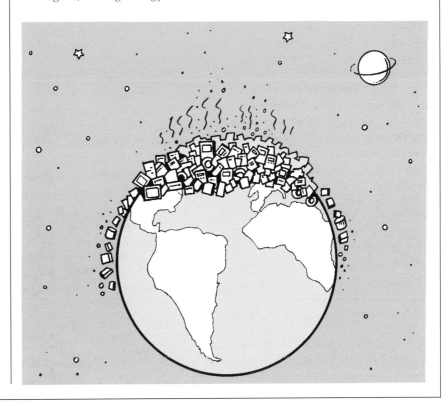

Recycling paper

Frequently asked questions

Does recycling 'save' trees?

The simple answer is no. This often-asked question implies that trees are under threat from the papermaker. In fact, because the papermaker generally uses the parts of the tree which other commercial users (such as builders and furniture makers) can't, it is not really appropriate to talk in terms of whole trees. The trees would still be harvested for these other commercial users.

Can we recycle for ever?

Paper can only be recycled approximately 4-6 times. Recycling prolongs the life of the wood pulp fibres used to make the paper, but the fibres themselves have a finite life, after which time their strength and quality regenerate. Recycling will never totally eradicate the need for virgin fibre.

Are we depleting the world's forests to meet our demand for paper?

The industry is very aware of the need to maintain the balance of natural resources. Great steps have been taken to ensure that forest levels are maintained and, as a consequence, more trees are being planted than are being harvested.

What about wildlife?

Every human activity is likely to affect wildlife. A conifer forest will support a different range of wild creatures. The forests now being

confederation of paper industries

felled were planted in the 1940s after the war had revealed the weakness of the UK being almost totally reliant on wood imports – but even they were far from devoid of life. These forests are being replaced today by woodlands shaped into the contours of the landscape, allowing generous fringes of different species. They are designed to provide habitats to encourage and expand the variety of wildlife.

Does recycling save vast amounts of energy?

Recycling and papermaking are both industrial processes that are energy intensive. In papermaking, the amount of energy needed depends on the process used. Mechanical pulping gives a high pulp yield but energy consumption is high. The chemical pulping process, which is used by most papermakers worldwide, gives a lower yield but, by burning wood residues, bark and pulping liquors, it provides steam and power which are often in excess of the amounts needed for the pulping and papermaking processes. In the recycling process waste paper often has to be de-inked which is energy consuming. Therefore, rather than

making a blanket statement that recycling saves energy, the levels of energy used or produced by each of the processes should be compared.

Landfill – recycling vs incineration?

Secondary uses for fibre include recycling and incineration for energy purposes. Our preferred route is recycling, although there are practical limits to the amount of waste paper that can be recycled. Incineration is a preferable option to landfill, but only once recycling requirements have been met.

Are paper mills polluting river water?

Stringent regulations are applied to the water which paper mills return to rivers. Papermakers recycle the water they use and it is very often the case that it goes back into the river cleaner than when it came out.

Do we use too much paper?

Paper has many natural attributes which make it a very attractive and versatile material to use. It comes from a naturally renewable source; it is recyclable and it is biodegradable. The amount of paper used has certainly increased, and this is partly due to developments in technology which have made paper suitable for many more purposes than originally imagined. Paper plays a very important role in everyday life. It can be easy to take it for granted but, like all the world's resources, paper should never be used thoughtlessly or wastefully. Common sense should prevail: don't use more paper than you need for the purpose; whenever possible recycle your paper once you have finished using it.

■ The above information is reprinted with kind permission from the Confederation of Paper Industries. For more information please visit www.paper.org.uk

© *Confederation of Paper Industries*

Making compost

Information from the Young People's Trust for the Environment

Why make compost?

The kitchen and garden waste that you throw in the bin ends up either being burned or being dumped in landfill sites.

BUT. . . it can be recycled and made into COMPOST!

What is compost?

Kitchen and garden waste is made up of organic matter. When you leave dead organic matter in a heap for a few days it will start to rot. Rotting is caused by bacteria, algae and fungi which eat the soft succulent bits of waste. All this eating makes them hot and encourages them to multiply. More heat and more bacteria mean that the organic matter is broken down more quickly. Organic matter is made up of anything that is, or was once alive.

Worms and beetles then eat the larger bits of waste. By the time all this is finished you are left with a dark brown crumbly mixture which is called COMPOST.

What can you compost?

- grass cuttings
- hay
- hedge trimmings
- paper (with no coloured ink)
- vegetable peelings
- straw

compost

FOR THE ENVIRONMENT

- tea bags
- leftover vegetables
- fruit.

Make a compost heap

How to start

A compost heap can just be a pile of kitchen scraps and garden waste. To speed up the rotting process, however, you can do several things.

Neat and tidy!

To keep your heap neat, you could make a wire mesh bin from chicken wire and four posts . . . you don't have to use chicken wire – wood or plastic sheets could be used instead.

Where to put it

Compost heaps like warmth, so, if you can, find a spot in your garden which is sheltered from the wind and in a sunny position. Remember compost heaps need to be damp but not waterlogged. Putting a piece of old carpet, wood or some newspapers on top helps keep out the rain.

Top tip!

If you line your compost bin with old cardboard or carpet it will keep it warm and damp – ideal conditions for bacteria to work in!

- The above information is reprinted with kind permission from the Young People's Trust for the Environment. For more information visit www.yptenc.org.uk or see page 41 for address details.

© Young People's Trust for the Environment

Composting

Composting is a natural process that occurs when degradable materials break down. Composting is easy and fun. All you need is a small outdoor space and a great recipe of air, water, carbon and nitrogen.

Why bother?

- About 33% of household waste can be composted.
- 78% of domestic waste ends up in landfill.
- Organic biodegradable waste is the main source of methane in landfill.
- Methane is one of the greenhouse gases responsible for global warming.
- Your garden benefits directly from recycling your own organic waste.
- Saves on the cost of buying compost and fertilisers.
- Helps preserve natural supplies of peat.
- Prevents pollution by avoiding sending organic waste to landfill.

For more info about WEN's composting projects, please take a look at the Local Food pages on the WEN website. You'll also find heaps of useful contacts in the Waste Links section.

Quick fact: Everyone's at it! Many local authorities now provide compost bins at knockdown prices. Give your local council a call and find out what's happening in your area.

- Information from the Women's Environmental Network. Visit www.wen.org.uk for more information or see page 41 for address details.

© WEN

English householders could win thousands by recycling

English householders could win thousands of pounds in cash or prizes for regularly recycling their waste, under various schemes being trialled from October 2005

The government today launched its bid to trial financial incentives to encourage householders around the country to recycle more of their waste.

Defra is funding 51 trials – worth a total of £3.5 million – to test various approaches for rewarding people who recycle their waste.

The government had said previously that up to £5 million would be available for the trials, but a Defra spokeswoman told letsrecycle.com it was decided after assessing bids that the full amount would not be needed.

Benefit

Schemes named today may benefit individual household recyclers themselves or local schools, communities and charities.

Among those being trialled include schemes using cash rewards, prizes and discount vouchers for shopping or local leisure facilities. There will also be recycling lotteries, league tables and scratch-card schemes trialled.

Householders in the Tees Valley could win a £3,000 holiday on their scheme, while recyclers in South Oxfordshire could win back their Council Tax. Communities in the Cornish borough of Restormel will compete in a recycling league to win £10,000.

The top prize in the draw being run by Exeter city council is a £12,000 car, which the council says will be an 'environmentally-friendly' model.

The results from the pilot studies announced today will provide an evidence base for future policy development in this area, Defra said, and will provide guidance to local authorities on best practice.

Momentum

Minister for environmental quality Ben Bradshaw said the new approaches were needed to achieve further substantial increases in recycling by actively engaging with the public.

Mr Bradshaw said: 'While there are millions of dedicated recyclers, there are still many families and people who have yet to start recycling regularly. We want to find new ways to encourage these people to start recycling and help regular recyclers by making it easier for them to fit recycling into their lives.'

> *Householders in the Tees Valley could win a £3,000 holiday on their recycling scheme, while recyclers in South Oxfordshire could win back their Council Tax*

Defra is hoping that financial 'carrots' like the schemes now on trial will be enough to make recycling a more common activity for householders, rather than resorting to 'sticks' to force householders to recycle – although this is something being considered as part of the government's waste strategy review. *6 October 2005*

■ The above information is reprinted with kind permission from Let's Recycle. For more visit www.letsrecycle.com or see page 41.

© Let's Recycle

Recycling tips

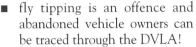

This article gives general advice on rubbish disposal and recycling. Always use your common sense and never place rubbish where it could cause harm. If you are at all unsure contact your Local Authority and speak to the recycling officer.

Asbestos

- a building and insulation material commonly used before the 1970s.
- the three main types of asbestos are white, blue and brown.
- can only cause harm if the fibres are inhaled.
- removal from buildings may disturb the fibres.

Advice: contact your local council before removal.

Batteries

- ordinary batteries contain no poisonous substances, they are safe to dispose of with everyday rubbish.
- rechargeable batteries contain hazardous metals, look for ones without mercury or cadmium.
- rechargeable batteries should be returned to the manufacturer where possible for disposal.
- some councils and garages provide facilities for recycling rechargeable batteries and lead acid car batteries.

Advice: buy rechargeable batteries, or appliances which use mains electricity.

Building rubbish

- includes: bricks, asphalt, glass, metals, plastics, soil, and wood.
- most of this waste is created by the construction industry.
- architectural salvage yards take some items for resale, and old bricks and timber can often be reused.

Advice: contact the Association for environmentally conscious building for advice.

Cars and vehicles (ELVs)

- over 1.8 million tonnes of old vehicles are thrown away in the UK each year.
- on average 75% of a vehicle is currently recycled.
- scrap merchants are able to recycle old vehicles.

- fly tipping is an offence and abandoned vehicle owners can be traced through the DVLA!

Advice: abandoned vehicles can be reported to your local council, who can also provide advice on disposal.

Chemicals, paint and oils

- chemicals are used every day in the home and garden, see the National Household Hazardous Waste Forum.
- DO NOT pour chemicals and oil down drains. They can pollute rivers.
- used engine oil can be recycled at most civic amenity sites.
- old paints and varnishes could be used by someone else – check the bank locator for community recycling schemes.
- plastic bottles which have contained household cleaners can also be recycled (check instructions on the bottle).

Advice: use environmentally friendly chemicals, most DIY stores stock them. Buy in bulk to reduce packaging.

Electronics

- some retailers take back old electrical items when delivering a new one.
- if your item still works safely, you could sell it. See *Yellow pages* for second-hand electrical shops.
- some charity shops will accept small electrical items.
- waste electronics can sometimes be recycled at council recycling sites.
- mobile phones can be recycled through phone retailers and charities.

Advice: try to repair broken items rather than throw them away. Buy durable items with long life cycles.

Furniture

- local charity shops, schools, and community groups can sometimes use unwanted items.
- please note that unwanted sofas

and chairs must have the kite mark to prove they meet British safety standards.

- most organisations will not take old beds for hygiene reasons.
- make sure all furniture is clean and in good repair before you donate it.

Advice: old furniture may be very useful to someone else! Donate unwanted items where possible.

Glass

- bottles and jars are usually separated by colour: brown, clear, and green.
- place in the correct colour bin (unless there is a mixed colour glass collection).
- wash out bottles and jars, remove caps and corks before recycling (avoid wasting water: use your washing-up water).
- light bulbs, Pyrex-type dishes, windowpanes etc. should not be put in glass banks.

Advice: reuse jars for storage, most supermarkets have glass banks, recycle alongside your weekly shop!

Medical waste

- dispose of medicines following either your doctor's or the manufacturer's instructions.
- care should be taken when disposing of needles and syringes.
- glass bottles and jars that have contained medicines can be recycled when they are empty.

Advice: if you find a syringe, use your common sense. If you can safely pick it up, then place it in a safe container and take it to the local police station.

Metals

- usually separated into: aluminium (drinks cans) – non-magnetic, and steel (food tins) – magnetic. Aerosols can be made from either.
- test by using a magnet.
- wash and squash cans before recycling. Only recycle clean aluminium foil. Never pierce or crush aerosols even when empty.
- only put empty aerosols in recycling schemes i.e. when you cannot get any more out by pressing the button.
- crisp wrappers (metallised plastic film) cannot be recycled. Metallised plastic springs back when scrunched.

Advice: contact Alupro for details of their Cash for Cans scheme and BAMA for any aerosol queries.

Paper and cardboard

- paper collection is usually separated into: newspapers, magazines, cardboard, and phone directories.
- unless specified, do not recycle catalogues, directories or envelopes which are gummed or glued together.
- juice and milk cartons cannot be recycled with ordinary paper as they are made up of several materials.
- some facilities provide mixed paper and card collection.

Advice: if you read newspapers, please recycle them after use. Alternatively, read news online. Set your printer to print double sided, buy recycled paper.

Plastic

- there are over 50 different types of plastics.
- if separate bins are provided it will usually be for:
  → HDPE – opaque bottles e.g. detergent bottles.
  → PVC – transparent bottles, an obvious seam running across the base e.g. mineral water bottles.
  → PET – transparent bottles, a hard moulded spot in the centre of the base e.g. fizzy drink bottles.
- some supermarkets have collection points for recycling carrier bags.

Advice: reuse bags or use a long-life carrier bag. Buy in bulk to reduce packaging.

Textiles

- old clothes, bedding, curtains, and blankets can be recycled on any high street at charity shops, but only donate clean usable items!
- some charities also have recycling bins for textiles.
- if you deposit shoes, tie them together so they don't get separated!

Advice: use any unrecyclable textiles as cloths around your home.

Timber / wood

- the disposal of wood in landfill sites causes problems as it is often bulky and decomposes slowly.
- scrap wood is collected at civic amenity sites for recycling.

Advice: many retailers now stock products made out of recycled wood or renewable wood sources – look on the FSC website for further information.

- The above information is reprinted with kind permission from Recycle More. Please visit their website at www.recycle-more.co.uk for more information or if you would like to write to them, please see page 41 for their address details.

© *Recycle More*

England nears 23% household recycling rate

English local authorities are now averaging a recycling rate of nearly 23%, according to unaudited figures published by Defra

The figures for 2004/05 suggest England is on track to meet the national target to recycle 25% of household waste by the end of this year (2005/06).

And, as Defra published the provisional figures today, it revealed that English local authorities are to be consulted on possible new statutory targets for 2007/08.

Recycling of household waste has doubled in the last four years, according to the provisional national estimate.

Performance around the country varies with, on average, residents in the North East recycling the least (16%) and people in the East of England recycling the most (29%). The greatest leap has been in the East Midlands, up 7% on last year to 27%.

Recycling figures showing the best councils for recycling are not to be published for some months, a Defra spokesperson told letsrecycle.com.

Local Environmental Quality Minister, Ben Bradshaw, said: 'These figures prove how much more people understand the importance of recycling compared to even just four years ago. There's no doubt we can be proud of our progress to date, but now it's time to build on that and start catching up with some of Europe's top recyclers.'

Consultation

In future, all local authorities will have to maintain and improve their recycling levels according to Defra, which is to publish a consultation shortly with proposals for new statutory performance standards for 2007/08.

Later in the year, the Government will consult on the format and level of any future performance standards for local authorities, as part of the review of its Waste Strategy 2000.

Recycling of household waste has doubled in the last four years, according to the provisional national estimate

Mr Bradshaw said: 'Local authorities will have to look at improving their recycling rates too. We want to be well on the way to our 2010 target of 30% of waste being recycled – sooner rather than later. Some of our European counterparts are doing it, some parts of this country are too, so it is not an unrealistic goal.'

In the meantime, Defra and the WRAP are continuing to work with local authorities and retailers to pilot and roll out new ways – from new technology at recycling 'bring' banks to financial incentives such as discount vouchers – to get people recycling more.

Incentives

A new multi-million pound pilot programme of local authority household incentives is due to start in October which will pilot, test and assess various approaches to incentivising household behaviour.

According to WRAP (the Waste & Resources Action Programme), which runs the Government's national Recycle Now campaign for England, every household could recycle up to 60% of its waste.

WRAP, one of the organisations tasked with improving the UK's resource efficiency, thinks the key to recycling success lies in maintaining the momentum.

'We all care about the environment in one way or another, and the great thing about recycling is that it's a really easy way in which we can each make an individual contribution,' explained Jennie Price, Chief Executive of WRAP.

'Local authorities have been working hard to boost awareness and to make it much easier for us to recycle. Nearly 80% of England's households now have doorstep recycling schemes – now we all need to make sure we use them.'
14 September 2005

■ Information from Let's Recycle. For more information visit www.letsrecycle.com or see page 41 for their address details.

© Let's Recycle

Coming to a bin near you. . .

The spy that tells how much rubbish you create. By Hugh Muir

Though he foresaw many ways in which Big Brother might watch us, even George Orwell never imagined that the authorities would keep a keen eye on your bin.

Residents of Croydon, south London, have been told that the microchips being inserted into their new wheely bins may well be adapted so that the council can judge whether they are producing too much rubbish.

If the technology suggests that they are, errant residents may be visited by officials bearing advice on how they might 'manage their rubbish more effectively'.

In the shorter term the microchips will be used to tell council officers how many of the borough's 100,000 bins the refuse collectors have emptied and how many have been missed.

While the move will be welcomed by environmentalists, it has sparked a row between the Labour-led council and Andrew Pelling, the Conservative who represents the area on the London assembly. He has tagged the microchips the 'spy in your bin'.

Mr Pelling said: 'The Stasi or the KGB could never have dreamed of getting a spying device in every household.'

He said the technology might yield information which could be misused.

'If, for example, computer hackers broke in to the system, they could see sudden reductions in waste in specific households, suggesting the owners were on holiday and the house vacant.'

But a spokesman for Croydon council said the fears were unjustified. 'What we don't want is people putting into their wheely bins tins and glass and paper and textiles, all of which could go into recycling bins. It is the way forward for waste management. We are not the only council thinking about it.'

He said the microchips would help the council fend off unwarranted criticism.

'We will have a confident response to customers who claim their bin may not have been emptied,' he added.

11 February 2005

© Guardian Newspapers Limited 2005

Top green facts

The top 10 'Green House' facts on recycling and the environment

1. Each year in Britain, we throw away 28 million tonnes of rubbish from our homes. This weighs the same as three and a half million double-decker buses. A queue of buses that long would go around the world one and a half times. (Source: The Green parent website.)
2. You can make 20 cans out of recycled material with the same amount of energy it takes to make one new one. (Source: The Green parent website.)
3. The UK produces 420 million tonnes of solid waste every year. That's the weight of five cars for each person every year. We only recycle 11% of it. (Source: The Green parent website.)
4. Incinerating 10,000 tonnes of waste creates one job, landfill the same amount of waste creates six jobs, but recycling the same 10,000 tonnes creates 36 jobs. (Source: The Green parent website.)
5. In just over a week, we produce enough rubbish to fill Wembley stadium. Over half of that waste can be recycled. (Source: DETR.)
6. Every tonne of paper recycled saves 17 trees. (Source: The Green parent website.)
7. Every year in the UK we use 13 billion steel cans which, if you placed them end to end, would stretch to the moon – three times! (Source: Steel Can Recycling Information Bureau.)
8. The energy saved from recycling one glass bottle is enough to power a light bulb for four hours. (Source: www.practicalhelp.org.uk.)
9. Recycling one plastic bottle can save the same amount of energy needed to power a 60-watt lightbulb for six hours. (Source: Recoup.)
10. We use over six billion glass bottles and jars each year. It would take you over three and a half thousand years to sing 'Six Billion Green Bottles'!

■ The above information is reprinted with kind permission from Recycle More. Visit www.recycle-more.co.uk or see page 41 for more information.

© Recycle More

Wasted energy?

Energy from waste must not be overlooked, say industry experts

The Institution of Civil Engineers (ICE) and the Renewable Energy Association (REA) today (21 April 2005) issued a joint report showing the huge potential for greater generation of energy from waste. The report, *Quantification of the Potential Energy from Residuals in the UK*, concludes that there is the opportunity for certain types of waste to produce up to 17% of electricity generated in the UK by 2020.

Almost 30 million tonnes of household rubbish were sent to landfill in England alone in 2003. The report states that more than half of this rubbish could be used to create enough power to light 2 million homes each year. A large majority of this waste is recognised in EU law as a source of renewable energy.

Peter Gerstrom, Chairman of ICE's Waste Management Board, commented, 'Instead of burying rubbish that is left after recycling it can be used to create electricity through a variety of measures. We are not generating enough renewable electricity, which means that the UK will not reach the EU Renewables Directive target of producing 10% of our electricity from renewable sources by 2010. We are even less likely to reach the next target of having 20% provided by renewables by 2020.

'Year on year the UK is producing more waste. Waste into energy will

RENEWABLE ENERGY ASSOCIATION

have environmental benefits by reducing the rubbish mountain. It also has the added bonus that recycling residual biodegradable waste in this way is an effective way of hitting the targets in the EU Landfill Directive.

'The UK should be taking the opportunity to harness this energy as this will boost our environmental performance by increasing our use of renewable power and reduce the UK's reliance on landfill. This will not happen in the current climate.

'The findings of this report should be of interest to the Government as the current DTI consultation *Renewables Obligation Review*, published 4 April 2005, has within its remit the opportunity to consider allowing producers of energy generated from waste to receive Renewable Obligation Certificates (ROCs).'

Gaynor Hartnell, Director of Policy at RPA, said, 'Many of our European neighbours excel at both

recycling and energy recovery. Producing energy from waste after recycling targets have been achieved is environmentally sound and will help us meet both our renewables targets and help us minimise the amount of waste going to landfill. It also helps with energy security, through reducing dependence on energy imports.'

Almost 30 million tonnes of household rubbish were sent to landfill in England alone in 2003... more than half of this could be used to create enough power to light 2 million homes each year

---

The UK should seek to limit the unsustainable option of landfill for Commercial and Industrial Waste, to bring us in line with the rest of Europe. This would encourage greater recycling and secure sufficient amounts of biodegradable waste to realise the 17% potential identified in the report. ICE and RPA are calling for government support to encourage the development of this energy resource.

Peter Gerstrom continues, 'It is patently not in the UK's interest to allow the energy, enough to power the population of Wales and Northern Ireland every year, to go to waste by being buried. Radical thinking about alternative energy, such as that highlighted in this report, is required to ensure the safety and diversity of the UK energy supply.'
21 April 2005

■ The above information is reprinted with kind permission from the Renewable Energy Association. Visit www.r-p-a.org.uk for more information.

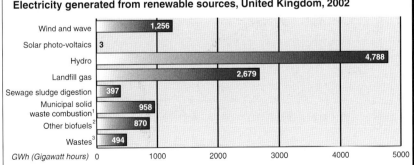

Electricity from renewable sources

Electricity generated from renewable sources, United Kingdom, 2002

Source	GWh
Wind and wave	1,256
Solar photo-voltaics	3
Hydro	4,788
Landfill gas	2,679
Sewage sludge digestion	397
Municipal solid waste combustion[1]	958
Other biofuels[2]	870
Wastes[3]	494

GWh (Gigawatt hours)   0   1000   2000   3000   4000   5000

1. Biodegradable part only.
2. Includes electricity from farm waste digestion, poultry litter combustion, meat and bone combustion, straw and short rotation coppice.
3. Non-biodgradeable part of municipal solid waste plus tyres.

Source: DTI. Crown copyright.

KEY FACTS

■ Waste or rubbish is what people throw away because they no longer need it or want it. Almost everything we do creates waste and as a society we are currently producing more waste than ever before. (page 1)

■ The process of using up the earth's natural resources to make products which we then throw away, sometimes a very short time later, is not 'sustainable' – in other words, it cannot continue indefinitely. (page 1)

■ About three million tonnes of plastic waste are produced in the UK each year, much of which is packaging (60 per cent). Once we have finished with plastic products, most of them are buried in the ground at landfill sites. As plastic is very durable and does not break down, this is where they will remain. (page 2)

■ Most rubbish collected from homes in the UK is buried in large holes in the ground (often old quarries) called landfill sites. Many of our current landfill sites are nearly full and we are rapidly running out of suitable land to create more. (page 3)

■ Most of the rubbish is produced by the countries who are part of the developed world, although they only make up about 5% of the world's population. The 'throwaway lifestyle' is not happening in every country. (page 5)

■ The volume of waste produced in the UK in one day is enough to fill Trafalgar Square to the top of Nelson's Column. (page 5)

■ It can cost a council between £6 and £19 per kilometre to sweep the streets. (page 8)

■ Fewer than 1% of the billions of plastic bags we use each year are recycled and the majority are used only once. (page 11)

■ The amount of fast food rubbish found on England's streets has risen by 50% in the last three years. (page 12)

■ Britain throws away nearly eight million nappies a day. (page 14)

■ 88% of household rubbish is disposed of in landfill sites. As the waste decomposes, it produces two dangerous substances, methane gas and leachate. (page 15)

■ In Britain, we each generate an average of 522 kg of waste per year. That's the same as 10 bags of sugar a week. (page 16)

■ Paper and card makes up 33% of the rubbish in our bins. (page 18)

■ It has been estimated that UK businesses lose up to 4.5% of annual turnover every year through avoidable waste. (page 18)

■ Fly tipping is the illegal deposit of any waste onto land. (page 19)

■ Recent figures show that waste is illegally dumped in England every 35 seconds and costs £100 a minute to clear. (page 20)

■ The amount of rubbish we produce has been escalating over the last 40 years. There has been a gradual change in shopping habits and people's attitudes to throwing things away. (page 22)

■ Approximately 3.5 billion glass bottles and jars, 1 billion plastic bottles, 2 billion aluminium cans and 2.5 billion steel cans were recycled by UK households in 2004. (page 25)

■ Recycling 120 aluminium cans saves enough energy to run a 3-bedroom house for one day. (page 26)

■ The energy saving from recycling one glass bottle will power a computer for 25 minutes. (page 28)

■ Products made from recycled materials can be indistinguishable from their counterparts manufactured from virgin materials. Aluminium, steel and glass can all be recycled repeatedly without a loss in quality. (page 29)

■ 80% of what we buy ends up being discarded within six months. (page 30)

■ Recycling paper reduces pressure on natural resources and uses 30-70% less energy than producing paper from virgin materials. (page 31)

■ Composting is a natural process that occurs when degradable materials break down. About 33% of household waste can be composted. (page 33)

■ The figures for 2004/05 suggest England is on track to meet the national target to recycle 25% of household waste by the end of this year (2005/06). (page 37)

■ Incinerating 10,000 tonnes of waste creates one job, landfill the same amount of waste creates six jobs, but recycling the same 10,000 tonnes creates 36 jobs. (page 38)

■ There is the opportunity for certain types of waste to produce up to 17% of electricity generated in the UK by 2020. (page 39)

ADDITIONAL RESOURCES

You might like to contact the following organisations for further information. Due to the increasing cost of postage, many organisations cannot respond to enquiries unless they receive a stamped, addressed envelope.

The Chartered Institution of Wastes Management
9 Saxon Court
St Peter's Gardens
NORTHAMPTON NN1 1SX
Tel: 01604 620426
Email: technical@ciwm.co.uk
Website: www.ciwm.co.uk
The Chartered Institution of Wastes Management represents over 5,000 professional people in the UK and overseas. The CIWA is dedicated to the protection and enhancement of the environment by the development of scientific, technical and management standards.

Department for Environment, Food and Rural Affairs (DEFRA)
Publications Department
Nobel House
17 Smith Square
LONDON SW1P 3JR
Tel: 020 7238 3000
Website: www.defra.gov.uk
Defra's remit is the pursuit of sustainable development – weaving together economic, social and environmental concerns.

ENCAMS (Environmental Campaigns)
Elizabeth House
The Pier
WIGAN WN3 4EX
Tel: 01942 824620
Email: information@encams.org
Website: www.encams.org
We are an environmental charity, which aims to achieve litter-free and sustainable environments by working with community groups, local authorities, businesses and other partners.

The Environment Agency (HQ)
Rio House
Waterside Drive
Aztec West
ALMONDSBURY
Bristol BS12 4UD
Tel: 08708 506506
Website:

www.environment-agency.gov.uk
The Environment Agency is a powerful and independent body which combines the regulation and management of land, air and water in England and Wales.

Friends of the Earth (FOE)
26-28 Underwood Street
LONDON N1 7JQ
Tel: 020 7490 1555
Email: info@foe.co.uk
Website: www.foe.co.uk
As an independent environmental group, Friends of the Earth publishes a comprehensive range of leaflets, books and in-depth briefings and reports.

Global Action Plan (GAP)
8 Fulwood Place
Gray's Inn
LONDON WC1V 6HG
Tel: 020 7405 5633
Email: all@globalactionplan.org.uk
Website:
www.globalactionplan.org.uk
Contact GAP for packs on: Action at Home, Action at Work and Action at School. Advice and support are available from trained local experts. Global Action Plan has volunteer groups around the country. Contact Global Action Plan for details and costs involved.

Let's Recycle
3/Downstream
1 London Bridge
LONDON SE1 9BG
Tel: 020 7785 6448
Email: news@letsrecycle.com
Website: www.letsrecycle.com
Fast, informed and authoritative – that's letsrecycle.com, the home of news and information for recyclers and all those involved in sustainable waste management in the UK today. letsrecycle.com is the UK's only independent dedicated website for businesses, local government and community groups involved in recycling and waste management.

Recycle More
Valpak Ltd
Stratford Business Park
Banbury Road
STRATFORD-UPON-AVON
Warwickshire CV36 9JT
Tel: 08450 682 572
Website: www.recycle-more.co.uk

Waste Watch
46-64 Leonard Street
LONDON EC2A 4JX
Tel: 020 7549 0300
Email: info@wastewatch.org.uk
Website: www.wastewatch.org.uk
www.recyclezone.org.uk
Waste Watch is a charity whose inspiration and values derive from a desire to protect the environment by ensuring the sustainable use and disposal of scarce resources, primarily by advocating waste reduction, reuse and recycling of materials.

Women's Environmental Network (WEN)
PO Box 30626
LONDON E1 1TZ
Tel: 020 7481 9004
Email: info@wen.org.uk
Website: www.wen.org.uk
Women's Environmental Network (WEN) is a UK-based membership charity that campaigns on environmental and health issues from a woman's perspective and educates, informs and empowers women and men who care about the environment.

Young People's Trust for the Environment
3 Walnut Tree Park
Walnut Tree Close
GUILDFORD
Surrey GU1 4TR
01483 539600
Email: info@yptenc.org.uk
Website: www.yptenc.org.uk
Works to educate young people in matters relating to the conservation of the world's wild places and natural resources.

INDEX

ACKNOWLEDGEMENTS

The publisher is grateful for permission to reproduce the following material.

While every care has been taken to trace and acknowledge copyright, the publisher tenders its apology for any accidental infringement or where copyright has proved untraceable. The publisher would be pleased to come to a suitable arrangement in any such case with the rightful owner.

Chapter One: The Problem of Waste

The problem with waste, © Waste Watch, *What a waste*, © Crown copyright is reproduced with the permission of Her Majesty's Stationery Office, *Waste words*, © Waste Watch, *A brief history of rubbish*, © Recycle More, *Waste facts*, © Women's Environmental Network, *What happens to the rubbish we produce?*, © Waste Watch, *Incineration and landfill*, © Friends of the Earth, *Quick facts about litter*, © EnCams, *Packaging waste*, © Chartered Institution of Wastes Management, *Food packaging*, © TheSite.org, *Pigeon population plump and rising*, © EnCams, *Study serves up nappy dilemma for parents*, © Crown copyright is reproduced with the permission of Her Majesty's Stationery Office, *The nappy debate*, © Women's Environmental Network, *The rubbish problem*, © Global Action Plan.

Chapter Two: Tackling Waste

Waste prevention and you, © Women's Environmental Network, *What can you do about waste?*, © Recycle More, *Waste in the workplace*, © Waste Watch, *Litter louts face new on-the-spot fines*, © Guardian Newspapers Ltd 2005, *Fly tipping*, © Chartered Institution of Wastes Management, *Dumping on Britain*, © Crown copyright is reproduced with the permission of Her Majesty's Stationery Office, *The Clean Neighbourhoods and Environment Act*, © Crown copyright is reproduced with the permission of Her Majesty's Stationery Office, *Recycling*, © Young People's Trust for the Environment, *Going green*, © British Glass, *Recyclable materials*, © Shanks Waste Management Ltd, *Is your brain full of rubbish?*, © Waste Watch, *Five reasons to recycle glass*, © British Glass, *Uses of recycled waste*, © Budget Pack, *Doorstep recycling survey*, © Friends of the Earth, *Should I bother recycling?*, © Guardian Newspapers Ltd 2005, *Recycle*, © Global Action Plan, *Recycling paper*, © Confederation of Paper Industries, *Making compost*, © Young People's Trust for the Environment, *Composting*, © Women's Environmental Network, *English householders could win thousands by recycling*, © letsrecycle.com, *Recycling tips*, © Recycle More, *England nears 23% household recycling rate*, © letsrecycle.com, *Coming to a bin near you*, © Guardian Newspapers Ltd 2005, *Top green facts*, © Recycle More, *Wasted energy?*, © Renewable Energy Association.

Photographs and illustrations:

Pages 1, 21: Pumpkin House; pages 3, 26, 34, 38: Simon Kneebone; pages 6, 37: Bev Aisbett; pages 8, 32, 36: Angelo Madrid; pages 9, 31, 35: Don Hatcher.

Craig Donnellan
Cambridge
January, 2006